The Ethology of Domestic Animals, 2nd Edition

An Introductory Text

DATE DUE

The authors would like to dedicate this book to a person whose devoted work in this area has contributed greatly to the progress of applied ethology. Long before it became widely accepted that understanding animal behaviour was an important aspect of understanding animal welfare, he actively promoted the development of applied ethology in veterinary medicine and animal science. The importance of his work for Swedish and European animal welfare development cannot be overestimated. The second edition is submitted at the same time as he celebrates his 80th birthday and, as this is written, his work continues.

To Ingvar Ekesbo

The Ethology of Domestic Animals, 2nd Edition

An Introductory Text

Per Jensen

IFM Biology
Linköping University
SE-58183 Linköping
Sweden

www.cabi.org

CABI is a trading name of CAB International

CABI Head Office	CABI North American Office
Nosworthy Way	875 Massachusetts Avenue
Wallingford	7th Floor
Oxfordshire OX10 8DE	Cambridge, MA 02139
UK	USA
Tel: +44 (0)1491 832111	Tel: +1 617 395 4056
Fax: +44 (0)1491 833508	Fax: +1 617 354 6875
Email: cabi@cabi.org	Email: cabi-nao@cabi.org
Web site: www.cabi.org	

A catalogue record for this book is available from the British Library, London, UK.

Library of Congress Cataloging-in-Publication Data
The ethology of domestic animals: an introductory text / edited by Per Jensen. – 2nd ed.
    p. cm.
  Includes bibliographical references and index.
  ISBN 978-1-84593-536-8 (alk. paper)
1. Livestock—Behavior. 2. Domestic animals—Behavior. I. Jensen, Per. II. Title.

SF756.7.E838 2009
636—dc22

                           2008048649

First edition 2002 ISBN 978-0-85199-602-8
Reprinted 2003, 2005, 2007 (twice)
Second edition 2009
Reprinted 2011

ISBN-13: 978-1-84593-536-8

Typeset by SPi, Pondicherry, India
Printed and bound in the UK by the MPG Books Group, Bodmin

The paper used for the text pages in this book is FSC certified. The FSC (Forest Stewardship Council) is an international network to promote responsible management of the world's forests.

Contents

Contributors

Melissa Bateson, *Division of Psychology, School of Biology and Psychology, Newcastle University, Henry Wellcome Building for Neuroecology, Framlington Place, Newcastle upon Tyne NE2 4HH, UK. Email: Melissa.Bateson@newcastle.ac.uk*

John Bradshaw, *Waltham Director of the Anthrozoology Institute, University of Bristol, Department of Clinical Veterinary Science, Langford BS40 5DU, UK. Email: J.W.S.Bradshaw@bristol.ac.uk*

Charlotte Burn, *Department of Veterinary Clinical Science, Royal Veterinary College, University of London, London NW1 0TU, UK.*

Cathy Dwyer, *Animal Behaviour and Welfare, Sustainable Livestock Systems Group, SAC, Bush Estate, Penicuik, Midlothian EH26 0PH, UK. Email: cathy.dwyer@sac.ac.uk*

David Fraser, *183 Macmillan–2357 Main Mall, Faculty of Land and Food Systems, University of British Columbia, Vancouver BC, Canada V6T 1Z4. Email: david.fraser@ubc.ca*

Laura Hänninen, *Research Centre for Animal Welfare, Department of Production Animal Medicine, PO Box 57, 00014 University of Helsinki, Finland.*

Per Jensen, *IFM Biology, Linköping University, SE581 83 Linköping, Sweden. Email: perje@ifm.liu.se*

Linda Keeling, *Department of Animal Environment and Health, Swedish University of Agricultural Sciences, Uppsala, Sweden. Email: linda.keeling@hmh.slu.se*

Naomi Latham, *Department of Zoology, University of Oxford OX1 3PS, UK.*

Georgia Mason, *Department of Animal & Poultry Science, University of Guelph, Guelph, Ontario, Canada N1G 2W1. Email: gmason@uoguelph.ca*

Joy Mench, *Department of Animal Science, University of California, Davis, One Shields Avenue, Davis, CA 95616-8521, USA. E-mail: jamench@ucdavis.edu*

Mike Mendl, *Department of Clinical Veterinary Science, University of Bristol, Langford House, Langford, UK. Email: mike.mendl@bristol.ac.uk*

Daniel Mills, *Dept of Biological Sciences, University of Lincoln, Riseholme Park, Lincoln LN2 2LG, UK. Email: dmills@lincoln.ac.uk*

Christine Nicol, *Department of Clinical Veterinary Science, University of Bristol, Langford House, Langford, UK. Email: c.j.nicol@bristol.ac.uk*

Sarah Redgate, *Department of Biological Sciences, University of Lincoln, Riseholme Park, Lincoln LN2 2LG, UK. Email: sredgate@lincoln.ac.uk*

Marek Špinka, *Ethology Group, Institute of Animal Science, 104 01 Prague – Uhříněves, Czech Republic. Email: spinka@vuzv.cz*

Cassandra Tucker, *Department of Animal Science, University of California, Davis, One Shields Avenue, Davis, CA 95616-8521, USA. Email: cbtucker@ucdavis.edu*

Anna Valros, *Research Centre for Animal Welfare, Department of Production Animal Medicine, PO Box 57, 00014 University of Helsinki, Finland. Email: anna.valros@helsinki.fi*

Susanne Waiblinger, *University of Veterinary Medicine, Institute of Animal Husbandry and Welfare, Veterinärplatz 1, Vienna A-1210, Austria. Email: Susanne.Waiblinger@vu-wien.ac.at*

Daniel M. Weary, *183 Macmillan–2357 Main Mall, Faculty of Land and Food Systems, University of British Columbia, Vancouver BC, Canada V6T 1Z4. Email: danweary@interchange.ubc.ca*

Deborah Wells, *School of Psychology, Queen's University Belfast, Belfast BT7 1NN, UK. Email: d.wells@qub.ac.uk*

Hanno Würbel, *Tierschutz und Ethologie, Institut für Veterinär-Physiologie, Justus Liebig Universität Giessen, Frankfurter Strasse 104, D-35392 Giessen, Germany. Email: hanno.wuerbel@vetmed.uni-giessen.de*

Contributors

Preface

Modern farm environments are profoundly different from the natural habitats of the ancestors of today's farm animals and, through genetic selection, both appearance and behaviour of the animals have changed. However, the legacy of these ancestors is still obvious, and some apparently bizarre actions can be understood only in the light of the evolutionary history of the species. On the other hand, some of the behaviour we can observe in animals on a modern farm or in a laboratory are not part of the normal, species-specific behaviour at all. They may even indicate that the animal is under stress and that its welfare is poor. Distinguishing between these possibilities is one important goal for applied ethology.

In this second edition of *The Ethology of Domestic Animals: An Introductory Text*, efforts have been made to expand the subject and to include novel and intriguing research findings. For example, the behaviour of dogs is now dealt with in a chapter of its own, and the topic of human–animal interactions is given an entirely new chapter. The issue of animal cognition, central to understanding welfare, has also received a more thorough examination. The readers we have had in mind when writing the book are, for example, students of animal science or veterinary medicine, or biology students taking introductory courses in animal behaviour or applied zoology. We have assumed that the readers will have some fundamental knowledge of basic biology – for example, physiology and genetics – and probably of modern animal husbandry, but also that this book will be one of their first texts in animal behaviour.

The authors of the book are largely new as compared with those who contributed to the first edition. They are all active in research, and experienced teachers. Hopefully, this will mean that the text has a suitable level and content. In order to make the text easier to read, the authors have reduced the references to a minimum, and these refer mainly to central publications that will provide a basis for further studies. Hence, not all cited experimental data are explicitly referenced, but can mostly be located via some of the other publications in the reference lists.

Per Jensen
Editor
Linköping, November 2008

PART I: BASIC ELEMENTS OF ANIMAL BEHAVIOUR

Editor's Introduction

The first eight chapters of this book introduce the basic concepts and the central subject matters required for a firm understanding of the biological bases of animal behaviour. The first chapter provides a historical background, which may help in understanding the questions that occupy contemporary ethology and its applied branches. In the second chapter, we approach the important issues of if and how behaviour is controlled by genes (the nature–nurture debate), and also what this means for understanding behavioural evolution. This chapter also describes the process of domestication, which is essential for understanding present-day domestic animals. The third chapter goes into some depth in describing how observable behaviour is a result of processes in the brain and throughout the body. The concept of motivation has a long history in ethology, and has proved essential for understanding the cognitive capacities and needs of animals in captivity, and this concept receives a detailed treatment in the fourth chapter. This leads naturally on to the fifth chapter, where learning and cognition are in focus. Here, considerable scientific advance has been made over the last decade, and this is introduced to the reader together with its relevance for animal welfare.

The sixth chapter moves to a more evolutionary and ecological approach to behaviour. Social and reproductive behaviour are both important elements of applied ethology, since animals are normally kept in groups, and are expected to reproduce. The seventh chapter attempts to bridge the gap between basic and applied ethology. It raises some of the central issues of contemporary applied ethology, such as how the behaviour of an animal can be used to assess its welfare and whether it is under stress. It also gives a broad and general outline of the most common and important behavioural disorders seen in captive environments. In the eighth and last chapter in the first part of the book, focus is shifted to the social interaction, with perhaps the most important counterpart of domesticated animals – the human. Extensive research has produced some exciting new aspects on this relationship, and this chapter provides a broad and up-to-date introduction to this.

1 The Study of Animal Behaviour and its Applications

P. JENSEN

1.1 Introduction

Most people have a clear conception of the meaning of the word 'behaviour', yet it is strikingly difficult to define it in a precise way. Since ethology is the science of animal behaviour, its causation and function, it is worthwhile to start with some consideration of what type of biological phenomena may be included in the concept. In its simplest forms, behaviour may be series of muscle contractions, perhaps performed in clear response to a specific stimulus, such as in the case of a reflex. However, at the other, extreme, end, we find enormously complex activities, such as birds migrating across the world, continuously assessing their directions and positions with the help of various cues from stars, landmarks and geomagneticism. It may not be obvious which stimuli actually trigger the onset of this behaviour. Indeed, a bird kept in a cage in a windowless room with constant light will show strong attempts to escape and move towards south at the right timing, without any relevant external cues at all.

We would use the word behaviour for both these extremes, and for many other activities in between in complexity. It will include all types of activities in which animals engage, such as locomotion, grooming, reproduction, caring for young, communication, etc. Behaviour may involve one individual reacting to a stimulus or a physiological change, but may also involve two individuals, each responding to the activities of the other. And why stop there? We would also call it behaviour when animals in a herd or an aggregation coordinate their activities or compete for resources with one another. No wonder ethology is such a complex science, when the phenomena we study are so disparate.

But how did it all begin and how has ethology developed into the science it is today? This chapter will provide a brief overview of some landmarks in history, and of the various fields into which the science has branched over recent decades. The field that will interest us most in this book is of course the applications of ethology to the study of animals utilized by man.

1.2 The History of Animal Behaviour Studies

No doubt, knowledge of animal behaviour must have been critical for the survival of early *Homo sapiens*. How could you construct a trap, or kill dangerous prey weighing several times your own weight, unless you have a genuine feeling for animal behaviour? So it should not be of any surprise that the earliest 'documents' available from humans – cave paintings of up to 30,000 years in age – are dominated by pictures of animals in various situations. Written, systematic observations and ideas about animal behaviour were published by Aristotle more than 300 years BC (Thorpe, 1979).

One of the first to write about animal behaviour in a modern fashion was the British zoologist John Ray. He published in 1676 a scientific text on the study of 'instinctive behaviour' in birds. He was astonished by the fact that birds, removed from their nests as young, would still build species-typical nests when adult. Ray was unable to explain the phenomenon, but noted the fact that very complex behaviour could develop without learning or practice. Almost 100 years later, French naturalists had an important influence on the development of the science. For example, Charles Georges Leroy, who was not actually a formally trained zoologist, published a book on intelligence and adaptation in animals. Leroy heavily criticized those philosophers who spent their time in their chambers, thinking about the world, rather than observing animals in their natural environments. Only then, he argued, would it be possible to fully appreciate the adaptive capacity and flexibility in the behaviour of animals (Thorpe, 1979). Figure 1.1 illustrates typical nest-building behaviour by a sow.

Another 100 years on, two important scientists deserve to be mentioned. The first is the British biologist Douglas Spalding, who published a series of papers on the relationship between instinct and experience. Spalding was way ahead of his time in experimental approaches. For example, he hatched eggs from hens by using the heat from a steaming kettle, in order to examine the development of the visual and acoustic senses without the influences of a mother hen (Thorpe, 1979). The second important scientist is none other than Charles Darwin.

Darwin is probably the one person who has had the most significant influence on the development of modern ethology – in fact on all modern biology. Most people know him as the father of the theory of evolution, which in itself of course is the foundation for any study of animal behaviour. However, he also approached the subject more directly, and his last published work in 1872, *The Expression of the Emotions in Man and Animals*, was probably the first, modern work on comparative ethology.

Fig. 1.1. A sow needs no prior experience to be able to construct an elaborate nest before farrowing. 'Instinctive behaviour' such as this fascinated the early behavioural scientists.

P. Jensen

1.3 Schools of Thought of the 20th Century

In the beginning of the 20th century, behavioural research grew fast. However, the development in the USA and Europe took different directions. American researchers were influenced by the behaviouristic approach, developed by people such as John B. Watson, and later by Burrhus Frederic Skinner. Their work was primarily focused on controlled experiments in laboratory environments, and their subject species *par préférence* were rats and mice. At the centre of their interest were the mechanisms of learning and acquisition of behaviour through reinforcement or punishment (Goodenough *et al.*, 1993). The behaviouristic research was concerned with finding general rules and principles of learning, and there was a strong belief that such rules were independent of context. Therefore, the evolutionary history of the study subjects or their ecological ways of life were regarded as irrelevant for the research.

In contrast, in Europe the development of the science was dominated by naturalistic biologists, who spent most of their time observing wild animals in nature. Birds and insects were favourite subjects, and these researchers were mostly interested in instinctive, innate and adaptive behaviour. One of the pioneers was Oskar Heinroth, who first started to use the term 'ethology' with the meaning we give it today (Thorpe, 1979). The naturalistic behavioural biologists shared an important approach with the behaviourists. They were not particularly interested in the mental processes or emotions that may be associated with behaviour. Such processes were often regarded as unavailable for scientific research, since they were not considered to be observable. Only much later has a scientific interest in mental processes emerged, something which will be dealt with more in later chapters of this book.

In the footsteps of Heinroth we meet two scientists whose influence over modern ethology cannot be overemphasized – Niko Tinbergen in The Netherlands and Britain, and Konrad Lorenz in Austria. Tinbergen developed a field methodology of high exactness. He designed experiments where details of the environments of free-living animals were altered and their behaviour could be recorded. He was a pioneer in experimental ethology (Dawkins *et al.*, 1991). Lorenz, on the other hand, did not go much into nature, but rather bred his experimental animals himself and kept many of them almost as pets. He rarely conducted elaborate experiments and was not prone to quantitative recordings. The strength of Lorenz was on the theoretical level: he formulated many of the fundamental ideas in ethology, and developed the first coherent theory of instinct and innate behaviour (Goodenough *et al.*, 1993).

Lorenz and Tinbergen definitely placed ethology on the solid ground of well-accepted sciences when they, together with the German researcher Karl von Frisch, were awarded the 1973 Nobel Prize in medicine and physiology.

1.4 Modern Approaches to Ethology

From the 1960s onwards, ethology developed into the science it is today. This was guided to a large extent by the research programme formulated by Tinbergen, and still generally accepted as the fundamentals of ethology (Tinbergen, 1963; Dawkins *et al.*, 1991). This programme is frequently referred to as 'Tinbergen's four questions', and the four aspects of behaviour that he felt were most important to ethology are:

1. What is the causation of the behaviour? The answer to this question refers to the immediate causes, such as which stimuli elicit or stimulate a behaviour, or which physiological variables, such as hormones, are important in the causation.
2. What is the function of the behaviour? In this case, the answer describes how the behaviour adds to the reproductive success, i.e. the fitness, of the animal. It is therefore involved with evolutionary aspects and consequences.
3. How does the behaviour develop during ontogeny? Studies on this question aim to describe the way a behaviour is modified by individual experiences.
4. How does the behaviour develop during phylogeny? This is clearly an evolutionary question, and usually calls for comparative studies of related species.

Whereas early ethology was mainly occupied by causation, ontogeny and phylogeny, the research during the 1960s and beyond became more and more focused on the functional question. Researchers have outlined new theories on how behaviour evolves through individual selection on the gene level, and have provided formal mathematical models for how the functional aspects of behaviour could be determined. The impact of this approach on contemporary animal behaviour science has been tremendous.

One aspect not covered by Tinbergen's questions was what animals perceive, feel and know in relation to their own behaviour. As mentioned earlier, this aspect of animal behaviour was largely considered to be inaccessible for science. However, other scientists have developed methods and concepts to allow investigation into this area. This has led to a new branch of the science, emerging in the 1970s, known as cognitive ethology (Bekoff, 2000) (the word cognition means subjective, mental processes – or thinking).

1.5 Applied Ethology

Even early in the development of ethology it was apparent that the new insights into the biology of behaviour could be of great value in understanding more of the behaviour of domestic animals. This branch of science saw a dramatic expansion as the debate on animal welfare in so-called factory farming (a concept coined by the influential writer Ruth Harrison) became aired in the 1960s. However, applied ethology is not only concerned with animal welfare. Let us look at a few areas of interest.

Welfare assessment

There is no doubt that the welfare of animals on farms, in zoos and laboratories dominates the interest of most researchers in this area. The problems may be formulated, for example, like this: most laying hens in the world are kept in cages made of wire mesh, with very little space available for the animals and almost no substrates for carrying out many of the species-typical behaviour patterns of poultry (see Fig. 1.2). So which are the most essential behaviour patterns for laying hens? Perhaps it is being able to dust bathe, or to perch during night or to perform nest building and laying eggs in a secluded area. All of these are typical poultry behaviours, and there may be others as well. How are the animals affected if they cannot behave like this? Can the activities be rated in any order of importance to the animals (Appleby *et al.*, 1993)?

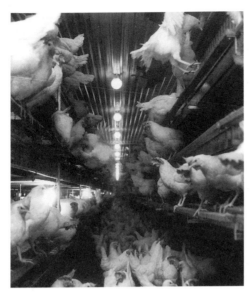

Fig. 1.2. Battery cages and floor housing systems both cause behavioural problems for laying hens. To estimate the relative importance of different behaviours to animals, thereby allowing better decisions regarding choice of housing systems, is one important goal of many researchers in applied ethology.

On the other hand, one common alternative to battery cages is a floor system with thousands of hens in one big group, sometimes with quite high stocking rates. In this situation, some unwanted behaviour (which can be present both in cages and in floor systems) may cause great harm to the animals, such as feather-pecking or cannibalism. So is it better for the hens to be in a situation where they can perform all the activities mentioned above, but where the social system may collapse and abnormal behaviour may spread widely (Hansen, 1994)?

Difficult questions such as these are important aspects of welfare assessment – only rarely do all indices point in the same direction. In this book, Chapter 7 will examine these aspects further and describe some of the methods researchers have developed to try to answer such questions.

Optimizing production

Farm animals are kept to produce food and other essentials for humans, and farmers need their enterprises to be profitable. It is therefore important that the difference between the value of what the animals produce (for example, amount of milk or meat) and the costs the farmer has for this production (for example, feed, investment and labour) is sufficiently high.

By taking knowledge of animal behaviour into account, such optimization may be easier to achieve. For example, animals may utilize their feed better if they are fed according to their species-specific feeding rhythm and in a social context adapted to the species (Nielsen et al., 1996). Social animals may eat more and digest the food better when all in a group are allowed to feed at the same time.

Social animals kept in individual housing systems may be poorer at transforming feed into valuable products. Likewise, husbandry routines applied at a biologically inadequate time may decrease the animals' production rate. Young piglets that are weaned from their mothers too early and in an abrupt manner show a decreased growth curve, and mixing of piglets after weaning may also have negative results on production (Algers *et al.*, 1990; Pajor *et al.*, 1991).

Behavioural control

The essence of keeping animals in captivity is to control their behaviour – by preventing them from escaping, to control their breeding and allowing them to adapt to the housing environment. Control is largely achieved by direct human actions, but also by the use of technical equipment.

A growing interest has been paid to the nature of man–animal interactions. For example, researchers have investigated how animals perceive humans and how they remember experiences with human behaviour. This may help farmers and others to interact more smoothly with their animals (Hemsworth and Barnett, 1987).

An increasing trend in animal farming is the use of technical innovation in animal husbandry. For example, group-housed pregnant sows are often fed from an electronic feeding station, which the animals to some extent are required to control themselves. The sows are equipped with transponders that allow them to open the feeding stations and obtain their individual feed rations. However, such systems must be carefully designed to avoid problems. For example, the social hierarchy of a group of sows may lead to some sows occupying the feed entrance, biting and wounding other animals and thereby destroying the functionality of the system. By means of ethological knowledge, technical equipment can be designed to work better for the animals (Broom *et al.*, 1995).

Behavioural disorders

Housing systems such as those described earlier, malfunctioning technical equipment or poor human management may all lead to various behavioural disorders. Aggression levels may become excessively high, dramatic behaviour such as cannibalism may develop and several other types of abnormal behaviour may also be seen (Lawrence and Rushen, 1993). This is not only the case for farm animals: many pets develop unwanted and abnormal behaviour, such as owner-directed aggression, uncontrolled urination and defecation or anxiety-like states.

The characterization and understanding of abnormal behaviour are a central aspect of applied ethology. Sometimes the behaviour can be remedied by behavioural therapies, such as enrichment of the home environment or stimulation of other behaviour. At other times research can provide insights that may help in avoiding the development of abnormal behaviours.

1.6　The Field of Applied Ethology

As should be clear from these accounts, ethology is a science that may offer many different sorts of applications in situations where humans utilize animals for various purposes.

P. Jensen

Whereas animal welfare assessment clearly dominates in the application of this science, it is by no means the only way in which knowledge of behaviour can be used.

Applied ethologists are normally concerned with all four of Tinbergen's questions. The causation and ontogeny of behaviour are essential aspects of understanding, for example, how abnormal behaviours develop and how they can be prevented. Phylogeny and function of behaviour are often less emphasized, but many studies have advanced our understanding of domestic animal behaviour greatly by considering how it can have evolved in previous generations and how it may have been affected by domestication (Fraser *et al.*, 1995). Experimental studies tend to dominate, but important scientific data have been made available through studies of domestic animals in wild-like conditions (which will become obvious in Chapters 9–16, where accounts are given of the normal behaviour of some important domestic species).

In the optimal situation, applied ethology research concerns all the fields outlined above. They are all interlinked: poor human–animal, or equipment–animal, interaction may cause poor welfare, which in turn leads to behavioural disorders and reduced production. Applied ethology is therefore an essential part of the correct keeping of animals. And last but not least: as will become obvious throughout this book, understanding the behaviour of domestic animals is a fascinating aspect of biology in its own right.

References

Algers, B., Jensen, P. and Steinwall, L. (1990) Behaviour and weight changes at weaning and regrouping of pigs in relation to teat quality. *Applied Animal Behaviour Science* 26, 143–155.

Appleby, M.C., Smith, S.F. and Hughes, B.O. (1993) Nesting, dust bathing and perching by laying hens in cages: effects of design on behaviour and welfare. *British Poultry Science* 34, 835–847.

Bekoff, M. (2000) Animal emotions: exploring passionate natures. *BioScience* 50, 861–870.

Broom, D.M., Mendl, M.T. and Zanella, A.J. (1995) A comparison of the welfare of sows in different housing conditions. *Animal Science* 61, 369–385.

Dawkins, M.S., Halliday, T.R. and Dawkins, R. (eds) (1991) *The Tinbergen Legacy*. Chapman & Hall, London.

Fraser, D., Kramer, D.L., Pajor, E.A. and Weary, D.M. (1995) Conflict and cooperation: sociobiological principles and the behaviour of pigs. *Applied Animal Behaviour Science* 44, 139–157.

Goodenough, J., McGuire, B. and Wallace, R.A. (1993) *Perspectives on Animal Behavior*. John Wiley & Sons, Inc., New York.

Hansen, I. (1994) Behavioural expression of laying hens in aviaries and cages: frequency, time budgets and facility utilisation. *British Poultry Science* 35, 491–508.

Hemsworth, P.H. and Barnett, J.L. (1987) Human–animal interactions. In: Price, E.O. (ed.) *Veterinary Clinics of North America: Food Animal Practice*, Vol. 3. W.B Saunders, Philadelphia, Pennsylvania, pp. 339–356.

Lawrence, A.B. and Rushen, J. (eds) (1993) *Stereotypic Animal Behaviour – Fundamentals and Applications to Welfare*. CAB International, Wallingford, UK.

Nielsen, B.L., Lawrence, A.B. and Whittemore, C.T. (1996) Feeding behaviour of growing pigs using single or multi-space feeders. *Applied Animal Behaviour Science* 47, 235–246.

Pajor, E.A., Fraser, D. and Kramer, D.L. (1991) Consumption of solid food by suckling pigs: individual variation and relation to weight gain. *Applied Animal Behaviour Science* 32, 139–156.

Thorpe, W.H. (1979) *The Origins and Rise of Ethology*. Heinemann Educational Books, London.

Tinbergen, N. (1963) On aims and methods of ethology. *Zeitschrift für Tierpsychologie* 20, 410–433.

2 Behaviour Genetics, Evolution and Domestication

P. JENSEN

2.1 Introduction

As we noted in the first chapter, people have been amazed for hundreds of years over the fact that animals are often able to perform extensive and complex, seemingly goal-directed behaviour with no prior learning possibilities. Birds may build elaborate nests and migrate to the typical wintering sites even if they are raised out of contact with other members of the same species. Darwin suggested that such 'instinctive' behaviour was somehow transferred from the parents without the need for any learning processes and, hence, 'instincts' would be suitable raw material for evolution. Behaviour would then evolve as adaptive traits just like any morphological or physiological character.

When Darwin formulated his ideas, he knew nothing about genes. The work of the 'father of genetics', Gregor Mendel, did not become known to the scientific world until several decades after Darwin's death, and the discovery of the chemical structure of DNA was still almost 100 years away. However, Darwin's suggestion was prophetic: we now know that genes, built up by DNA, contain the chemical instructions necessary for the development of behaviour in an animal, and that evolution modifies the frequency of genes over generations. Evolution therefore moulds the behaviour of species and individuals. In this chapter, we will examine some of the evidence for this, and its implications.

The focus of this book is on domesticated animals, so of course we will need to ask how domestication has come about, and how animals have been changed as a consequence of becoming domesticated.

2.2 To What Degree is Behaviour Genetically Inherited?

Anyone interested in dogs and dog breeds would have no problem in accepting the idea that even complex behaviour is genetically inherited from the parents. Retrievers have puppies that are, on the whole, more inclined to retrieve and carry things around than other breeds, while Border collies have puppies with a strong tendency to herd. Some breeds are more noted for being aggressive, while others are generally viewed as friendly or docile. Observations such as these are suggestive, but they could perhaps also be explained by, for example, the offspring learning a specific behaviour from their parents, or by the fact that some types of owners tend to keep some types of dogs, and therefore may affect their behaviour differently. So how can we tell whether and to what extent behaviour is inherited genetically?

One of the more famous examples of a clear genetical influence on behaviour comes from studies on lovebirds, *Agapornis* spp. (Dilger, 1962). One species, Fischer's lovebird, carries nesting material (for example, strips of paper) one piece at a time in the beak. The closely related peach-faced lovebird tucks the strips in between the rump feathers and is therefore able to carry more nesting material at each flight. Hybrids between the two show a poorly functioning mixture of the two behaviours; they attempt to tuck material between the feathers, fail to let go of it, pull it out again and then start the sequence over again. After several months of practising, this behaviour can become at least partly successful, in that the birds manage to transport some material back to the nest site, but not in a manner typical of any of the parents. This strange behaviour of the hybrids is consistent with an intermediate (non-dominant) inheritance pattern of one or more alleles (an allele is one variant of a gene that is present in the population; hence every gene may appear in several allelles).

Another source of evidence is the weight of scientific experimentation, which shows that the tendency to behave in a certain manner has been artificially selected over generations. Mostly, this leads to individuals of the selected lines being more and more inclined to behave according to the selection criteria, indicating a strong genetic basis for the behaviour. For example, in fruit flies (*Drosophila melanogaster*), mating time may vary considerably between individuals. In a classical experiment (see Fig. 2.1), 100 fruit flies of both sexes were placed in a chamber and the ten fastest and the ten slowest to mate were determined. They were allowed to breed separately, and the procedure was repeated in each new generation. After 25 generations, mating time differed between populations by about 30 min (Manning, 1961). Similarly, the tendency for positive versus negative geotaxis, i.e. to move towards or against gravitation on a vertical surface, varies between fruit flies, and selection for the strength of this tendency produced rapid changes in the population over several generations (Hirsch, 1967).

Experiments such as these indicate that even complex behavioural responses can be controlled by genes. When there is a quantitative response, behaviour becomes gradually faster or slower, or a tendency is gradually strengthened or weakened, it is most likely that several genes interact in a quantitative fashion in producing the observed behaviour phenotype. Later, we will examine some cases where few, or even single, genes affect the behavioural output.

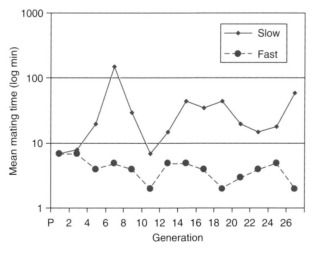

Fig. 2.1. Results of a selection experiment where fruit flies were selected for long or short mating time (modified after Manning, 1961).

Do studies of insects have any relevance for applied ethology, which is mostly concerned with mammals and birds? Yes, genetic control is universal among animals. Not only are the tools – DNA, RNA, etc. – identical, but the way in which the control is exerted is largely the same (we will deal with the mechanisms later in this chapter). Moreover, the sequencing of the complete, or almost complete, genomes of organisms ranging from yeast, via the nematode *Caenorhabditis elegans* and fruit flies to chickens and various mammals – including humans – has revealed large similarities in structures and functions of many genes. Hence, genetic instructions for behaviour may sometimes be very similar across species and phyla.

There is of course also much direct evidence of genetically determined behavioural differences among domestic animals. For example, laboratory mice were selected for their tendency to build nests. After 15 generations, one line collected about 50 g of cotton for nesting material, compared with 5–10 g in the other line (Lynch, 1980). As another example, laying hens sometimes develop a problematic behaviour called feather-pecking, where they damage the feathers of flockmates. The tendency to develop this behaviour differs between strains, and genetic selection against the behaviour can cause it to decrease over several generations (Kjaer and Sorensen, 1997). This indicates a genetic basis for this abnormal behaviour.

Genetic selection of desired behaviour, or against undesired behaviour, is therefore also an important means for improving animal welfare. For example, levels of fearfulness have been reduced by selection in both poultry and quail (Jones and Hocking, 1999). Fearfulness has also been reduced in some fur animals, along with undesirable behaviours, like chewing and destroying the fur in mink (Malmkvist and Hansens, 2001).

How can there be genes controlling the onset of abnormal behaviour such as feather-pecking? This is probably not how it works. We must always remember that, even though selection is performed on a well-defined trait, the observed responses can often be explained by indirect effects. The reduction of feather-pecking by selection has been suggested to be indirectly caused by a general decrease in foraging activity, which reduces the overall tendency of the birds to peck (Klein *et al.*, 2000). Therefore, in this case we cannot be certain that there is a direct genetical control of feather-pecking as such – in fact it may seem more logical to assume that the effect is often indirect.

2.3 Genetic Versus Environmental Influence on Behaviour

Sometimes, people misinterpret the fact that genes contain instructions for the behaviour of an animal. This misinterpretation is often referred to as genetic determinism, the belief that, if there is a genetic control, the behaviour of an individual will be inflexible and determined from the point of fertilization. The following example illustrates nicely why this is wrong.

Rats will normally learn easily to run through a maze, in order to reach a goal box containing some food. With increasing numbers of trials, they will make fewer and fewer errors, in the sense that they become less and less likely to run into blind alleys. However, there is considerable individual variation in how easily this task is accomplished. In a famous experiment, rats were selected depending on how fast they learned a specific maze. In only a few generations, the selected lines had separated, with almost no overlap, into a line of so-called 'bright' rats and another of so-called 'dull rats' (Tryon, 1940). These lines could be preserved over generations and generations, and the offspring would show the same pattern in learning ability as their parents.

One possible interpretation of this experiment is that learning capacity is controlled by genetic factors. Once it is known from which population an animal descends, whether its father and mother are dull or bright, it would be possible to predict accurately how well this animal would manage a maze-learning test. However, this conclusion is oversimplified, as shown by the following, later, study.

Rats from the same populations were raised in three different environments: (i) the standard laboratory environment in which the populations had been maintained for generations; (ii) an impoverished environment without bedding material or any other interesting stimuli; and (iii) an enriched environment, where different substrates for manipulation and stimulation were added to the cage (Cooper and Zubek, 1958). When these animals were tested in the maze, those reared in the standard environment showed the same differences between lines as expected: the 'dull' line had considerably greater difficulties in mastering the problem.

However, when the behaviour of the rats from the two lines reared in a restricted environment was compared, an interesting result emerged. The differences between the strains had disappeared, and they both performed as poorly as the dull rats reared in the standard environment. Furthermore, there were also no differences between the strains when they had been reared in enriched environments, but this time they both performed as well as the bright strain from standard cages (see Fig. 2.2).

These results show how careful we must be in inferring deterministic genetic control over behaviour when a genetic correlation has been demonstrated. Clearly,

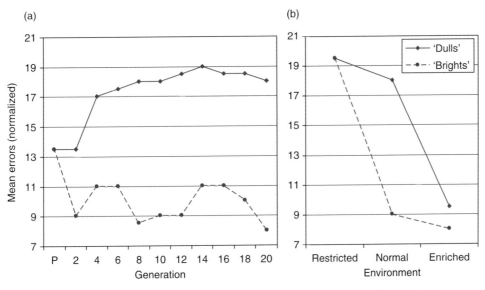

Fig. 2.2. (a) Results of selection for maze-learning capacity in rats, modified after Tryon (1940); (b) results of a maze-learning experiment where individuals from each of the selected lines were raised in either the standard laboratory environment, an impoverished laboratory environment or an enriched one (modified after Cooper and Zubek, 1958).

genes supply the organism with the necessary basis for a particular behaviour – the limbs, muscles, nerves and sensory organs, and a central nervous system – but any behaviour is likely to develop in interaction with the environment in which the animal lives. Rather than considering genes as determinants of behaviour, we should consider genetic traits to be predispositions, which bias animals towards certain reactions and developmental pathways.

This approach also helps us understand the possible indirect genetical influence on feather-pecking, described earlier. If there is a genetic predisposition to peck at different substrates, feather-pecking may develop when the environment lacks necessary pecking stimuli. Hens in a rich environment may therefore never develop this behavioural disturbance, although the same individuals would do so under poorer conditions.

2.4   Single Gene Influences

Since behaviours are complex traits, involving identification of stimuli acting on different sensory channels, central nervous processing of information and the concerted actions of groups of muscles, it may seem implausible that single genes would be able to control the expression of behaviour. Nevertheless, there are several examples of such effects of single – or at least of very few – genes.

In a comprehensive study of behavioural genetics, two breeds of dogs with very divergent behaviour, the basenji and the cocker spaniel, were crossed (Scott and Fuller, 1965). By producing backcrosses and using standard genetic analysis methods, it was possible to suggest simple genetical control mechanisms for rather complex behaviour patterns. For example, struggling when being restrained – more pronounced in the basenji – was consistent with inheritance by one single allele, showing no dominance. The tendency to bark was suggested to be inherited through two genes, with dominant alleles for a low stimulus threshold (the basenji, of course, is famous for having a very high barking threshold, i.e. they rarely bark).

In fruit flies there are many examples of effects of separate loci on complex behaviour. One mutant strain, called 'dunce', has a mutation in a single gene on the X chromosome. These flies cannot learn to avoid an odour when trained in an experiment that allows normal flies to connect the odour with an electric shock (Quinn *et al.*, 1974).

2.5   How do Genes Affect Behaviour?

Studies like those summarized above demonstrate clearly that genes have a strong influence over behaviour. But what does a gene actually do to affect behaviour, trapped as it is inside the nucleus of the cell? How is it possible that mutations in one or a few genes can change complex patterns of behaviour? DNA is a fantastic molecule for storing information, but the information in a gene is used only for providing the instructions for how and when a protein should be produced. What is the link between proteins and behaviour?

Since genes carry the codes for synthesis of proteins, they thereby regulate important aspects of metabolism, hormone concentrations, etc. Of course, the direct interaction between the environment and the animal, including its behaviour, is mostly activated via the nervous system. However, the construction of neurons into an entire nervous system and the design and sensitivity of receptors and sense organs are all processes that are regulated by genes during the stages of embryo and maturation. In the adult animal, levels of hormones, neurotransmittors and other substances that are extremely important for behavioural control are regulated by genes.

The detailed pathway from the synthesis of a specific protein to an observed behaviour is long and complex, and we have actually only started to understand some of it. For example, the 'dunce' fruit flies, discussed above, have a defect gene that is normally responsible for the production or activity of an enzyme called cyclic AMP phosphodiesterase, which degrades cyclic AMP (a common intermediate molecule in cell metabolism). So either the enzyme or the cyclic AMP is involved in olfactory learning. Once pathways such as these are more fully understood, they will no doubt provide deep insights into the biology of behaviour.

Even if we have seen examples of single-gene effects on behaviour, a certain behaviour is not usually controlled by a single gene. However, a single gene can have a crucial role for the normal appearance of a certain behaviour. Single mutations may alter the behaviour drastically, for example by changing the development of important structures (Huntingford, 1984). One mutation in the nematode *Caenorhabditis elegans* changes the myosin filaments in parts of the body and produces uncoordinated movements. Another mutation affects the structure of chemoreceptors in the head and changes the receptivity of the worm to chemical stimuli. The slow-mating fruit flies that we discussed earlier were suggested to have an altered production of juvenile hormone as a result of selection of genes for this.

2.6 Localizing Genes

Even if it is clear that genes predispose animals to specific types of behaviour, the mechanisms whereby this is happening are still largely unknown, particularly in vertebrates. Usually, we do not even know exactly which genes are involved in controlling a behaviour, and we therefore cannot understand what these genes might do to exert their effects. With the rapid development of genetics, identifying these genes and their functions is likely to be an important area of research in the immediate future.

A first step towards identifying genes tied to specific behavioural traits is to localize the position of the genes on the chromosomes. The position can sometimes be determined because a behaviour is inherited in close correlation with some other trait, for which the gene and its location is known – what geneticists refer to as linkage. It could, for example, be a colour pattern or some other observable trait. One can then conclude that the behaviourally linked gene must be located close to the gene for the other trait.

With modern gene technology, new possibilities for gene localization have emerged. One common method is to use quantitative trait loci (QTL) analysis. With this analysis, the genomes of animals are mapped with respect to the occurrence of specific DNA-markers, usually non-coding sequences of base pairs that vary between individuals in specific chromosomal locations. The correlation between a particular phenotypic trait, for example a behaviour, and the occurrence of different markers can then be assessed (sounds simple, but requires many animals, a lot of laboratory

Red junglefowl   White leghorn laying hen

Fig. 2.3. When junglefowl are crossed with modern laying hens and the F1 generation (upper photographs) is subsequently intercrossed, the resulting F2 generation (lower photographs) exhibits a large genetic and phenotypic variation. Such breeding experiments are important tools in QTL studies, i.e. experiments designed to localize the chromosomal loci of genes controlling specific traits.

wizardry and advanced statistical computing). When a certain trait has been found to be correlated with the occurrence of a certain marker, one can imply that the gene that affects the trait is located close to that marker. By using the increasingly detailed gene maps that are available for more and more species (humans, for example), relevant genes in that chromosome region can be identified, and their products studied (see Fig. 2.3).

QTL analyses have identified chromosomal regions influencing such variable aspects of behaviour as the tendency of honey bees to sting, the cyclicity of activity in mice, preference for alcohol in mice and the hyperactivity syndrome of rats.

Methods of identifying the actual genes and their products are developing rapidly, and in the next few years we will most probably see a tremendous increase in our understanding of behavioural genetics. One important source is going to be so-called 'knockout animals'. These are strains of animals where specific genes have been rendered silent. The development of such an animal, compared with normal individuals, therefore tells us important things about the function of the knocked-out genes. As an example, mice lacking the gene for oxytocin (a hormone involved in many different processes such as parturition and milk ejection) not only – as expected – lack the capacity to eject milk, but also show a reduction in aggressive behaviour (Crawley, 1999).

2.7 Evolution of Behaviour

Now that we have seen how behaviour depends on genetic predispositions, we have two of the necessary keys to understand how behaviour can have developed during evolution. It is nowadays considered basic biological knowledge that animals are products of a long evolutionary history, whereby their anatomy and physiology have been adaptively shaped to what we see today. But is that also the case for behaviour?

In order for any trait to be modified by evolution, three principles are required, which can be deduced from Darwin's original writings:

1. The principle of variation. This states that a trait must vary between the individuals of a population. If all individuals are identical, no evolution of the trait is possible.
2. The principle of genetic inheritance. This principle requires that some of the variation in the population must be of genetic origin. It is not necessary that the trait is genetically determined, only that genes have some influence over the phenotypical expression of the trait. It follows from this principle that, on average, when a trait is genetically inherited, individuals resemble their parents more than they resemble other randomly chosen individuals of the population with respect to this trait. Furthermore, the closer the relationship between any two individuals, the greater the resemblance.
3. The principle of natural selection. According to this, some variants of the trait must influence various reproductive abilities of individuals. If the reproduction capacity is enhanced, the trait will increase in frequency over generations and, if it is reduced, the frequency will decrease.

In order for evolution to modify any trait, all these principles must be fulfilled simultaneously. In fact, when they are all fulfilled, evolution is bound to happen. In the case of behaviour, we have already seen that there is often a large variation within populations (for example, in the case of maze-running ability in rats), and that this variation is often partly caused by genetic differences between individuals. What remains is to show that this genetically based influence produces variable reproductive success.

Indeed, most contemporary evolutionary research in behaviour is devoted to examining the reproductive advantages of having a certain behaviour rather than any other possible alternative. Obtaining and defending a territory may increase the chance of acquiring a mate and reproducing, and territory quality and size are often found to be closely related to reproductive success. Searching for food in a manner that causes minimal exposure to predators may seem an obvious example, and different food-searching patterns do have effects on the efficiency of reproduction. Several decades of intense research have produced innumerable similar examples of behaviour patterns that follow the principle of natural selection, and this is often covered in the subject termed behavioural ecology (Krebs and Davies, 1991).

2.8 Tracing the Evolutionary History of Behaviour

Behavioural archaeology would be a rather fruitless field of science, but occasionally fossils may actually carry some information about the behaviour of animals long extinct. For example, close examination of the skeleton and feather structure of the oldest bird fossil, *Archaeopteryx*, has led to the suggestion that bird flight developed through gliding from tree branches, and running and leaping after prey on the ground (Alcock, 2001) (although other interpretations have also been suggested). Mostly, however, we have to rely on comparisons of extant animals.

The method applied in this type of studies is to take advantage of phylogenetic trees, which may be deduced from traits other than behavioural, and then map the behaviour seen in closely related species on to this tree. For example, in one study of courtship displays in manakins – a group of small, fruit-eating tropical birds – a phylogenetic tree was constructed based on similiarities between species in the structure of the vocal apparatus (syrinx). The male courtship behaviour of the different species was then examined in detail. There were 44 behavioural elements that occurred in at least one of the 28 examined species. By mapping the species to the phylogenetic tree based on how many of the behavioural traits they shared, it was possible to deduce a possible history of how present-day courtship has evolved from simpler behaviour in different ancestors. Mostly, evolution has led to more elements being included in the repertoire but, in some cases, behavioural patterns have been lost during evolutionary development (Prum, 1990).

Sometimes, the mechanisms forming behavioural differences between species have been elucidated by comparative studies. The greenish warbler, *Phylloscopus trochiloides*, exists in several subspecies throughout Russia, Siberia and China. In its western range, the subspecies interbreed readily. However, over time, subpopulations have been separated by the Tibetan Plateau and, where the populations are again juxtaposed, they are in effect two different species. They produce completely different songs, and the females do not recognize the songs of the other population as species-specific (Irwin *et al.*, 2001). Geographical separation and female partner choice have driven the formation of a new warbler species.

2.9 Modification and Ritualization

The reconstruction of evolutionary history of behavioural traits leaves us with some important lessons. First, evolution works only by selection of variants of traits already present in an animal. This means that evolution, of course, cannot invent any new behaviour in the face of a 'need'. Hence, the reason why an animal uses a particular behaviour for a certain purpose is often to be found in the history of the species.

Inevitably, all behaviours we see today are therefore modified versions of behaviour patterns that may have served very different purposes in ancestral forms. A behaviour which serves a certain function becomes slightly modified and may then serve a slightly new function and, as time and generations pass, a new behaviour has developed. We can verify this causative chain since the original behaviour may be conserved in closely related species, with its original function still intact.

A special case of this is the evolution of animal signals by the process of ritualization. One of the most quoted examples of a ritualized display is that of the peacock's courtship display. The enormous and colourful tail is spread like a fan, vibrating, at the same time as the male is bowing towards the female. A likely evolutionary history looks like this: ancestors attracted females by pecking at feed items on the ground and emitting a special call – this is still the common courtship of one close relative, domestic poultry. These movements have become exaggerated, as in pheasants, and finally become unrelated to feed presentation, as in the peacock. The tail has developed and increased in size at the same time. A behaviour that originally served as food presentation has now become the courtship of peacocks (Schenkel, 1956). Ritualization is

Fig. 2.4. Domestic cockerels court females in the same way as junglefowl. One important element is 'titbitting', where the male emits a special call at the same time as he pecks at the ground, sometimes towards food particles. This has, by ritualization, developed into the spectacular peacock courtship display.

therefore the process by which a certain behaviour evolves into a signal by becoming exaggerated and losing its original function.

Ritualization, combined with exaggeration of anatomical traits (such as the male peacock tail (see Fig. 2.4)), is also driven by preferences of the females, which, since Darwin's time, have been referred to as sexual selection. A trait – for example a colour, or an ornament – may evolve because it indicates important qualities in the male. Perhaps only males that are genetically healthy and resistant to parasites can grow large tail feathers – the female that prefers to mate with large-tailed males may then produce offspring carrying the favourable genes (Alcock, 2001).

2.10   The Function of Behaviour

Another important lesson we have learnt from evolutionary biology is that behaviour does not evolve for 'the good of the species'. Since all evolution can do is to work on variation between individuals, there is simply no mechanism around which can take the view of a group of animals, or a species (but there are noteworthy exceptions, where the group may be an efficient unit of selection; see Chapter 6, this volume, for more details). When we look for the function of any particular behaviour, we have always therefore to consider the benefits of the individual. That benefit ultimately has to be measured in reproduction. The reproductive success of an individual is also referred to as its fitness.

A behaviour has not only potential fitness benefits for the performer, but also costs. It may consume valuable energy, which could otherwise be used for reproduction, or expose the animal to predators. Once we realize that a behaviour has both costs and benefits, it becomes obvious that evolution will select the behaviour that maximizes the difference between fitness benefits and costs. This is called the optimal behaviour (see Fig. 2.5).

For example, large territories are usually better, since they provide more food and other resources. Consequently, females often prefer males holding large territories. Should a male therefore always attempt to defend as big a territory as possible? No, big territories require much more energy and time to defend, and leave less for reproductive efforts. Therefore, males should aim for optimal territories that maximize the benefits received from attracting more females in relation to the costs. Here is another example: how long should an animal continue to search for food in the same place rather than moving to another, perhaps more profitable, site? The answer is that the animal should weigh up the information available regarding how depleted the present food patch is (number of food items ingested per time unit), how profitable another average patch is likely to be in the habitat and how long it will take to find another patch. Then it should choose to leave the present patch when net energy intake over time is maximized. This general idea is central to what is called optimal foraging theory (Krebs and Davies, 1991).

It may seem implausible that animals would be able to perform complicated calculations like this. Nevertheless, many studies have shown that animals generally behave in an optimal manner (Krebs and Davies, 1991). But, of course, they do not perform the mathematics implied in the theory. Evolution has selected animals with the highest fitness, and they are the ones most likely to thrive. If we could ask the animal why it leaves a patch after a certain time, it would not be able to tell us the evolutionary reason. Could it speak, it would probably say that 'it felt like leaving now'. In the technical jargon of behavioural ecology, the animal has a behavioural strategy that causes it to react to stimuli as if it were consciously maximizing fitness by optimizing its behaviour.

The same reasoning can be used to understand social behaviour. What is best for an animal to do in a social contest (for example, to attack or retreat) depends strongly on what the opponent does or is likely to do. Even though it may superficially appear beneficial always to attack and drive away intruders from a food source, some further

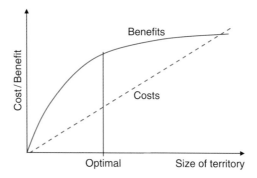

Fig. 2.5. Costs and benefits of a territory both increase as a function of the size of the territory, as is shown here. The functions are not identical – cost often increases linearly or with an increasing function, whereas benefits follow a decreasing function. This is because an animal cannot gain infinitely from larger territories. There are, for example, time constraints. Evolution has favoured animals choosing to defend territory of optimal size, i.e. where the difference between benefits and costs is greatest.

P. Jensen

thinking shows us that this is not necessarily so. Since the tendency to attack depends on genes, a population where all individuals always attack each other may be very unstable. A mutation that avoids fighting may gain a lot in fitness because the individual is not wounded, and the mutated gene may therefore spread quickly throughout the population. The optimal social strategy is therefore not always easy to prescribe without the use of formal mathematical modelling. The strategy that, on average, confers the largest benefit to the individuals of a population is usually referred to as the evolutionarily stable strategy, or ESS (Krebs and Davies, 1991).

2.11 Domestication: an Evolutionary Case History

Domestication, the process whereby an animal is transformed from a life in the wild to a life under some control of humans, is one of the most dramatic evolutionary processes accessible to scientific investigation. It is often treated in the literature as a cultural phenomenon, and the story often goes something like this (Clutton-Brock, 1999): when man became sedentary and agriculture began, some 10,000–15,000 years ago, those animals available for exploitation were tamed by man so they could provide food, clothing, etc. This led to the first domesticated species – dogs, goats and sheep, later followed by the main agricultural species, such as pigs, cattle and chickens. The reason why the story is often told in this way is that bones with typical domesticated traits (shortened legs, compressed skulls, etc.) have been found in excavations of early agricultural sites. According to this 'standard model' of domestication, each domestic species emerged from one single ancestral species and from one single domestication event, or from a few events. However, a biological approach to the subject reveals a somewhat different story.

Modern DNA technology has provided new tools for examining a variety of questions. For example, differences between animals in sequences of DNA in the mitochondria (mtDNA, which is inherited only from the mother) reveal their degree of relatedness and also the time that has passed since the animals became reproductively isolated from each other. Examinations of such data show that, for example, dogs and pigs dissociated from wolves and wild boars much earlier than archaeological evidence has suggested; perhaps as early as over 50,000 years ago, wolves that were later to develop into the domestic dog became reproductively isolated from the line that led to today's wild wolves. Domestication of pigs seems to have happened independently in Europe and Asia and, again, reproductive isolation of the populations occurred much earlier than indicated by archaeology (Giuffra et al., 2000). Independent events of horse domestication have been suggested from similar types of data (see Chapter 10).

This is consistent with a theory of domestication claiming that at least some animals in effect 'chose to be domesticated' (Budyansky, 1992), in the following sense. Some populations seem to have been adapted to take advantage of a newly arising, highly productive yet ephemeral habitat. The new species, *Homo sapiens*, emerging on the arena of life between 100,000 and 200,000 years ago, may have provided a potentially rich and fruitful niche for such animals. They may have enjoyed large fitness benefits from associating with man, simultaneously increasing human fitness by being relatively easy to hunt or by guarding against danger (in the case of wolves). It appears that these populations became reproductively isolated from the rest of the species long before domestication is thought to have begun. Perhaps domestication took a leap when humans became sedentary and agriculture started, and perhaps what archaeologists discover may be the results of the onset of active breeding. However, domestication as an evolutionary process seems to have been going on for perhaps five or ten times longer than this.

2.12 Which Animals were Domesticated?

One feature of domestication that requires a biological explanation is the similarity in some specific behavioural and ecological traits of animals. Not every animal species appears suitable for domestication: from a systematic perspective, there is a large bias among domesticated animals towards ungulate mammals and gallinaceous birds. From a behavioural aspect, there is a huge dominance of gregarious omnivores or herbivores with no strong mating bonds.

These behavioural traits (which are abundant among ungulates and gallinaceous birds) may be crucial for a successful domestication (Price, 1997). Social life allows many animals to live together in the human setting, and predisposes for hierarchical systems, where humans can more easily adopt the role of a dominant group leader. Feeding habits that do not compete with humans would be essential for successful cohabitation. Weak bonds between mates would make selective breeding much easier.

The first wave of domestication comprised the major present-day farm animals, and animals such as the dog and the horse. When agriculture was established in large parts of the world, many of these animals were already present and, from a biological perspective, could be considered domesticated. For thousands of years few new species were added (but of course a variety of breeds within each species developed). During more recent times, a second wave occurred, and a number of new species were domesticated, but there is no doubt that this was a process controlled completely by man, and dictated by specific needs or wishes. This process gave us fur animals (mink, foxes, raccoon dogs, chinchillas, etc.), laboratory animals (mainly mice and rats) and several new meat producers (buffaloes, ostriches, salmon). The animals domesticated in this second wave do not necessarily show the typical traits outlined earlier. For example, mink are solitary, territorial animals and foxes have strong mating bonds.

2.13 Behavioural Effects of Domestication

Since domestication involves the genetic modification of a population of animals, we would expect domestic animals to differ in a number of traits from their wild ancestors. Typical morphological changes in colour (more white and spotted individuals), shape (size and relative leg length) and function (less pronounced seasonality in reproduction) may lead us to expect that there will be pronounced behavioural differences as well. However, most research has found only subtle differences between domestic and wild animals (Price, 1997; Fig. 2.6). Most typically, these differences can be attributed to modified stimulus thresholds, causing some behaviour patterns to become more common and others to be rarer during domestication. No new behaviours seem to have been added to the behavioural repertoire of any domestic species, and few of the ancestral behaviour patterns have disappeared completely.

(a)   (b)

Fig. 2.6. Domestic pigs differ in appearance from wild boars, but their behaviour is very similar. (a) Sow foraging in the forest; (b) wild boar.

Hence, pigs kept for generations in restricted indoor housing systems still build elaborate farrowing nests if released into a forest, and laying hens kept in battery cages will attempt to perch high up at night if given the slightest opportunity.

The threshold differences that have occurred are sometimes caused by active selection by man. For example, most dogs have been bred to bark in response to very low stimulation, whereas the basenji, used for sneak-hunting in Africa, has been bred for the opposite. Other behavioural changes may be a result of an evolutionary adaptation. For example, energetically costly behavioural strategies appear in some cases to have been reduced in pigs and poultry (Gustafsson *et al.*, 1999; Schütz and Jensen, 2001). Furthermore, domestic animals are generally less fearful towards humans and more socially tolerant towards conspecifics.

Although there are behavioural differences between wild and domestic animals, it is clear that these are not as large as we sometimes tend to believe. It is also frequently suggested that domestic animals are less responsive to their environment than are wild animals, and even that they are 'more stupid'. However, detailed studies of the behaviour of domestic animals in natural conditions reveal that their behaviour is very similar to that of their ancestors. The fact that they may often develop abnormal behaviour, and even pathologies, when they are prevented from performing a normal behaviour indicates a strong responsiveness towards the environment in which they are kept. Regardless of whether we think that welfare concerns are central to animal husbandry, or whether we are guided by concerns over animals' productivity, we need to remember that the behaviour of domestic animals is controlled by genetic mechanisms shaped over thousands and thousands of generations of evolution in the wild, and only slightly altered during domestication. The evolutionary history and adaptations of their ancestors and the natural behaviour of the present-day animals are therefore important pieces of information if we want to understand the animals we keep for our benefit.

References

Alcock, J. (2001) *Animal Behaviour – an Evolutionary Approach*. Sinauer Associates Inc., Sunderland, Massachusetts.
Budyansky, S. (1992) *The Covenant of the Wild – Why Animals Chose Domestication*. William Morrow and Co., Inc., New York.

Clutton-Brock, J. (1999) *A Natural History of Domesticated Mammals*. Cambridge University Press, Cambridge, UK.

Cooper, R.M. and Zubek, J.P. (1958) Effects of enriched and restricted early environments on the learning ability of bright and dull rats. *Canadian Journal of Psychology* 12, 159–164.

Crawley, J.N. (1999) Behavioral phenotyping of transgenic and knockout mice. In: Jones, B.C. and Mormède, P. (eds) *Neurobehavioral Genetics – Methods and Applications*. CRC Press, Washington, DC, pp. 105–119.

Dilger, W.C. (1962) The behavior of lovebirds. *Scientific American* 206, 88–98.

Giuffra, E., Kijas, J.M.H., Amarger, V., Carlborg, Ö., Jeon, J.-T. and Andersson, L. (2000) The origin of the domestic pig: independent domestication and subsequent introgression. *Genetics* 154, 1785–1791.

Gustafsson, M., Jensen, P., Jonge, F.H.D., Schuurman, T. and de Jonge, F.H. (1999) Domestication effects on foraging strategies in pigs (*Sus scrofa*). *Applied Animal Behaviour Science* 62, 305–317.

Hirsch, J. (1967) *Behaviour Genetic Analysis*. McGraw Hill, New York.

Huntingford, F. (1984) *The Study of Animal Behaviour*. Chapman and Hall, London.

Irwin, D.E., Bensch, S. and Price, T.D. (2001) Speciation in a ring. *Nature* 409, 333–337.

Jones, R.B. and Hocking, P.M. (1999) Genetic selection for poultry behaviour: big bad wolf or friend in need? *Animal Welfare* 8, 343–359.

Kjaer, J.P. and Sorensen, P. (1997) Feather pecking behaviour in white leghorns, a genetic study. *British Poultry Science* 38, 333–341.

Klein, T., Zeltner, E. and Huber-Eicher, B. (2000) Are genetic differences in foraging behaviour of laying hen chicks paralleled by hybrid-specific differences in feather pecking? *Applied Animal Behaviour Science* 70, 143–155.

Krebs, J.R. and Davies, D.B. (1991) *Behavioural Ecology: an Evolutionary Approach*. Blackwell Scientific Publications, Oxford, UK.

Lynch, C.B. (1980) Response to divergent selection for nesting behaviour in *Mus musculus*. *Genetics* 96, 757–765.

Malmkvist, J. and Hansens, S.W. (2001) The welfare of farmed mink (*Mustela vison*) in relation to behavioural selection: a review. *Animal Welfare* 10, 41–52.

Manning, A. (1961) The effects of artificial selection for mating speed in *Drosophila melanogaster*. *Animal Behaviour* 16, 108–113.

Price, E.O. (1997) Behavioural genetics and the process of animal domestication. In: Grandin, T. (ed.) *Genetics and the Behaviour of Domestic Animals*. Academic Press, Lake Charles, Louisiana, pp. 31–65.

Prum, R.O. (1990) Phylogenetic analysis of the evolution of display behavior in the neotropical manakins (Aves: Pipridae). *Ethology* 84, 202–231.

Quinn, W.G., Harris, W.A. and Benzer, S. (1974) Conditioned behavior in *Drosophila melanogaster*. *Proceedings of the National Academy of Sciences of the United States of America* 71, 708–712.

Schenkel, R. (1956) Zur Deutung der Phasianidenbalz. *Ornithologische Beobachtungen* 53, 182.

Schütz, K. and Jensen, P. (2001) Effects of resource allocation on behavioural strategies: a comparison of red junglefowl (*Gallus gallus*) and two domesticated breeds of poultry. *Ethology* 107, 753–765.

Scott, J.P. and Fuller, J.L. (1965) *Genetics and the Social Behavior of the Dog*. University of Chicago Press, Chicago, Illinois.

Tryon, R.C. (1940) Studies in individual differences in maze ability VII. The specific components of maze ability and a general theory of psychological components. *Journal of Comparative Physiology and Psychology* 30, 283–335.

Behaviour and Physiology

A. Valros and L. Hänninen

3.1 Introduction

The behaviour of an animal reflects a complex series of underlying physiological events, both neurological and endocrinological. The motivation to perform certain behaviours is affected by both acute changes in the homeostatic balance of the individual – such as the reaction to a sudden change in the environmental temperature – and long-term internal or external processes – such as the reproductive state of the individual or the nutritional level of the environment. The physiology of an animal has an amazing capacity to adapt itself to changing situations. In addition, as the mechanisms behind the shown behaviour are extremely complex and often involve several interrelated mechanisms, it is often easier to measure and interpret the behaviour of an animal than to understand the underlying physiological mechanisms.

Behaviour can be considered a black box – showing us what is going on without always revealing why. For example, studies by Janczak *et al.* (2006, 2007) have shown that similar changes in the behaviour of the chicken, reflecting adaptations to prenatal stress, could be achieved both by injecting corticosterone into fertilized eggs and by stressing the mother hens during the laying period. However, the two methods of 'stressing' the eggs did not seem to be governed by the same physiological mechanisms, as the eggs from the stressed hens failed to show an increased level of corticosterone. However, in spite of the complex nature of the physiological mechanisms governing an individual's behaviour, we do need a basic understanding of these if we are to understand the reasons for the animal's behaviour.

3.2 Central Regulation of Behaviour

Hypothalamus and hypophysis: the 'master gland'

One of the most central brain regions governing the motivational state, and thus the behaviour of animals, is the hypothalamus. The hypothalamus is situated at the base of the brain and is extremely important in regulating several homeostatic mechanisms and related emotional reactions, such as body temperature, ingestion of food and water, sexual behaviour and rage (Eckert, 1988). The hypothalamus regulates the autonomic nervous system and links the nervous system to the endocrine system, thus being at the top of the endocrine hierarchy. At the basis of the hypothalamus a small, but very important, appendage can be found, i.e. the hypophysis, or the pituitary gland, which has been named the 'master gland' for its important role in governing peripheral endocrine functions (see Fig. 3.1).

The hypothalamus regulates the hypophysis largely through neurosecretory cells, i.e. nerve cells that produce hormones and secrete them into the bloodstream. These cells react to input from different parts of the body, related to such mechanisms as temperature regulation, osmoregulation and sexual cycles. Some of these cells reach throughout the posterior lobe of the hypophysis, or the neurohypophysis, which is really more like an extension of the hypothalamus, and release so-called neurohypophyseal hormones directly into the bloodstream. These hormones thus affect peripheral target tissues directly and include antidiuretic hormone (ADH), also called vasopressin, which increases water reabsorption in the kidney, and oxytocin, which effects smooth muscular contractions and is positively related to both social and maternal behaviour of mammals, as well as to milk production and 'let-down'.

The other type of neurosecretory cells in the hypothalamus release peptide hypothalamic releasing and inhibiting hormones into the bloodstream. From the hypothalamus the blood is directed into a capillary network in the anterior lobe of the hypophysis, or the adenohypophysis. Here, the hypothalamic releasing and inhibiting hormones attach to cell receptors and thus regulate the release of hormones from the non-neural endocrine cells in the adenohypophysis. Three of these adenohypophyseal hormones are regulated by both stimulatory and inhibitory hypothalamic hormones, and act directly on non-endocrine target tissues (see Table. 3.1). These include growth hormone (GH), prolactin (PL) and melanocyte-stimulating hormone (MSH). GH is important for growth, milk production and immunology and PL for milk production and maternal behaviour. MSH is actually secreted from the pars intermedia, a section between the anterior and posterior lobes of the hypothalamus, and regulates pigment formation.

The release of the remaining four adenohypophyseal hormones, i.e. adenocorticotropin (ACTH), thyrotropin (TSH), luteinizing hormone (LH) and follicle-stimulating hormone (FSH), is stimulated by a hypothalamic releasing hormone, while their inhibition

Table 3.1. Hormones secreted from the hypothalamus and the adenohypophysis and their target tissues or target glands. RH, releasing hormone; RIH, release-inhibiting hormone.

Hypothalamic hormones (and abbreviations)	Adenohypophyseal hormones (and abbreviations)	Target tissues or target glands (and their hormones)
Growth hormone RH (GH-RH) Growth hormone RIH (GH-RIH)	Growth hormone (GH)	All tissues
Prolactin RH (P-RH) Prolactin RIH (P-RIH)	Prolactin (PL)	Mammary tissue
Melanocyte-stimulating hormone RH (MSH-RH) Melanocyte-stimulating hormone RIH (MSH-RIH)	Melanocyte-stimulating hormone (MSH)	Pigment cells
Corticotropin RH (C-RH)	Adenocorticotropin (ACTH)	Adrenal cortex (corticosteroids)
Thyrotropin RH (T-RH)	Thyrotropin (TSH)	Thyroid gland (thyroid hormones)
Luteinizing hormone RH (LH-RH)	Luteinizing hormone (LH)	Gonads (sex steroids)
Follicle-stimulating hormone RH (FSH-RH)	Follicle-stimulating hormone (FSH)	Gonads (sex steroids)

A. Valros and L. Hänninen

is dependent on a negative feedback system. This means that their further release is inhibited via the hypothalamus, by the hormones released from their target glands into the blood circulation. All these hormones are important for the behaviour of animals: FSH and LH regulate reproductive behaviour by affecting the secretion of sex steroids; ACTH is a central part of the stress reaction, as it governs the secretion of corticosteroids; and TSH stimulates the release of thyroid hormones, which, in turn, increase the oxygen consumption and the metabolic rate, thereby also affecting thermoregulation. In addition, thyroid hormones are very important for the normal development and growth of individuals.

Emotion and consciousness

The limbic system involves several closely interacting parts of the brain. The main parts include the hypothalamus, the hippocampus and the amygdala, all situated just beneath the cerebrum, or the forebrain (see Fig. 3.1). In addition, several other structures are often considered part of the limbic system, such as the orbitofrontal cortex, which is important for decision making, and the nucleus accumbens, which is involved in producing feelings of pleasure and in addiction. The limbic system reacts to external and internal pleasant or unpleasant stimuli and coordinates changes needed to maintain homeostasis.

It has been shown that the amygdala is one of the key brain structures involved in emotional processes in both humans and animals (Cardinal *et al.*, 2002). However, in order for a stimulus to cause a conscious feeling or emotion it is believed that other parts of the brain, such as the forebrain, need to be involved: in order for a stimulus to result in an emotion it needs to be processed and related to expectations and associations, thus involving higher-order brain functions (Ressler, 2004). In many cases, animals appear to show similar brain reactions to stimuli as do humans and exhibit similar brain structures as those proved to be related to human experiences. But, as Dawkins (2006) states, even though there are considerable data on how animals of different species react to stimuli and which brain areas and physiological processes are involved in these reactions, there is still no commonly accepted answer to the question of which physiological and neurological processes result in consciousness and in sentience, and which species have these abilities.

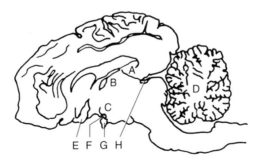

Fig. 3.1. Cross-section of a pig brain. A, hippocampus; B, hypothalamus; C, hypophysis; D, cerebellum; E, nucleus accumbens; F, suprachiasmatic nucleus; G, optic chiasm; H, pineal gland.

Decision making, the reward system and neurotransmission

One of the most important mechanisms steering the behaviour of an animal is the reward system. The simplified mechanism of this system is that an animal reacts differently to external or internal stimuli causing pleasant (rewards) or unpleasant (punishment) sensations (Ressler, 2004). Thus, a reward is something that an animal will work towards achieving, such as food or comfort, and a punishment something an animal will work to avoid, such as a predator or discomfort. The limbic system is involved in regulating the process of acting on potential rewards and punishments.

The amygdala is an important component of the reward system, which is largely based on dopaminergic neurons, i.e. neurons with dopamine as their transmitter substance. Dopamine is a monoamine, a similar compound to other established neurotransmitters such as serotonin and noradrenalin. Dopaminergic neurons are involved in several behaviour-related regulative systems of the brain, regulating, for example, functions related to cognition, motivation, sleep and mood. The dopaminergic system has also been connected to the performance of stereotypies in, e.g., hens.

The way an individual processes stimuli to make decisions can be simplified, as seen in Fig. 3.2. Sensory input is processed by different areas of the brain, such as the amygdala and the prefrontal cortex, thus allowing the individual to evaluate the input in relation to the expected reward value and prior experiences. At least in higher vertebrates, such as primates, different reaction possibilities are then evaluated and compared and a voluntary choice is made on how to act in the current situation (Opris and Bruce, 2005).

Serotonin, or 5-hydroxytryptamine (5-HT), is another important monoamine involved in the regulation of behaviour. Serotonin has been studied extensively in connection with human mental disorders, such as depression and compulsive–obsessive disorders. In animals, serotonin has been connected with the level of fear and anxiety, stress reactivity and stereotypic behaviour. By feeding tryptophan, an amino acid used by the body to synthesize serotonin, the behavioural reactions of animals can be altered. For example, tryptophan feeding or administration has been shown to reduce fear and anxiety in fur animals and reduce stress reactivity in pigs, probably due to an increase in brain serotonin levels.

Also, neuropeptides are involved in the reward system, producing pleasure- and euphoria-like states in animals. Endogenous opioids, such as endorphins, found mainly in the hypophysis, and encephalins, found throughout the nervous system, have been studied extensively in this context. These neuropeptides are produced by the central nervous system and bind to opioid receptors. They cause a reduction in the sense of pain, i.e. have an analgesic effect, and their levels rise in response to

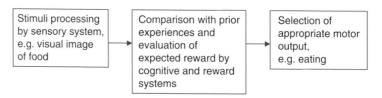

Fig. 3.2. Schematic description of the decision-making system (modified from Opris and Bruce, 2005).

A. Valros and L. Hänninen

pleasurable events, such as eating. This system is also involved in drug addictions in humans: narcotic opiates such as opium and morphine bind to the same opioid receptors, thus causing a pleasurable feeling. Repeated use of narcotic opiates causes a change in the neuronal metabolism, resulting in the individual feeling extreme discomfort until more drugs are administered. This is called addiction. In animals, it has been shown that endogenous opioids are related to the level of stress response, and to the performance of stereotypies.

Control of circadian rhythms

The body's biological clock is a hierarchical system, governed by the suprachiasmatic nucleus (SCN), which is the main regulator for biological rhythms. The SCN is situated in the hypothalamus, directly above the optic chiasma (see Fig. 3.1). Even though most of the peripheral tissues and organs also generate their own circadian rhythms, their molecular clock is entrained by the SCN. In mammals, the SCN is regulated mainly by light via the retinohypothalamic tract (RHT), which is a pathway of photoreceptive cells from the retina to the SCN. In mammals, the hormone melatonin is secreted mainly during darkness from the pineal gland, a small, conical structure in the central part of the brain (see Fig. 3.1). Secretion is driven by the SCN, and can be thought of as a neuroendocrine transducer of the light–dark cycle. Melatonin is an important modulator of circadian rhythms; it promotes sleep and can induce phase shifts of biological rhythms. Other factors that synchronize the circadian system include, for example, nutrition, hormone feedback mechanisms, activity and social cues (Buijs *et al.*, 2003).

Hormonal fluctuation and rest–activity or sleep rhythms are examples of biological rhythms. These rhythms are classified as either circadian, with a cycle of approximately 24 h, or ultradian, with a cycle under 24 h. Many of the hormones mentioned above have circadian rhythms, such as ACTH, cortisol, GH and prolactin; also, body temperature follows a circadian cycle. This, of course has a marked effect on the individual's behaviour, alertness and cognitive output.

Examples of circadian rhythms involving both physiology and behaviour include the connection between sleep and hormones such as GH and glucocorticoids. Sleep is a physiological regulator for GH in laboratory animals and humans, and sleep is associated with GH secretion also in sheep. Typically, sleep onset stimulates GH secretion. The secretion of GH increases during sleep independent of the circadian sleeping cycle, and sleep deprivation diminishes the GH release. In humans and many of the animal species studied so far, blood glucocorticoids are highest when waking up in the subjective morning, and lowest in the subjective evening and early part of the sleep period. Thus, sleep normally sets in when the corticotropic activity is quiescent, at least in humans and laboratory rodents. Cortisol begins to rise a few hours before the usual waking time. The cortisol rhythm is very stable, but sleep deprivation changes the pulse amplitude.

3.3 Sleep

All animals benefit from restricting their different activities to times of the day when optimal environmental conditions exist – for example, when it is not too dangerous to eat or sleep. Thus, during evolution, each species has developed its unique sleeping method, level and rhythm. Primates often exhibit mono- or biphasic sleeping rhythms while several other species have a polyphasic rhythm, sleeping all through the day. Typical examples of polyphasic, but also short, sleepers are our domestic ruminants and horses. Birds, on the other hand, have developed a very unique sleeping style – many of them, such as domestic hens, are capable of sleeping unihemispheric sleep, i.e. one cerebral hemisphere sleeps while the other remains awake. When one hemisphere is asleep, the opposite, contralateral, eye is closed while the other eye is open. This type of a sleep makes it possible for birds to keep an eye on predators, even when asleep. Birds are capable of controlling which brain side is asleep according to predator pressure.

Sleep can be defined electrophysiologically as consisting of two main phases: rapid eye movement sleep (REM), in earlier literature also called paradoxical sleep or active sleep; and non-rapid eye movement sleep (NREM), also called quiet sleep, orthodoxical sleep or slow wave sleep. The stimulus threshold for waking from REM sleep is higher than that for NREM sleep, and thus a long REM sleep period can be a threat to survival for prey species other than those rodent species that sleep in nests.

The sleep cycle consists of one or several REM and NREM phases, and the cycle length is species specific. The cycles are short in farm animals, as is usual in prey animal species. For example, calves sleep in several 5 min periods throughout the 24 h day (Hänninen, 2007; Fig. 3.3). Total sleep duration varies a lot between species (see Fig. 3.4). These differences between species depend on several factors, such as time spent eating, available food sources, digestion and rumination, and ecological niche. Grazing animals, for example, are assumed to sleep less, as they need more time to consume large amounts of low-calorie foods (Siegel, 2005).

Fig. 3.3. Young calves sleep for approximately 20% of the day in short bouts throughout the 24 h period. The photograph shows a calf with its head to the side and neck relaxed, indicating REM sleep (image courtesy of Patrik Pesonius).

A. Valros and L. Hänninen

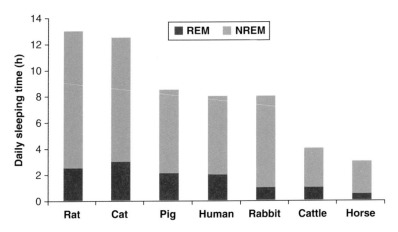

Fig. 3.4. Total daily sleeping time (REM and NREM sleep) in hours for various animal species. Data according to Zepelin *et al.* (2005), except for cattle (from Ruckebusch, 1972).

In mature mammals and birds, the sleep phase usually starts with NREM and deepens during the REM phase. The young of many terrestrial animal species sleep more and have more REM sleep than do older animals. Sleep is essential for the development of the brain, and REM sleep is connected to the early developmental phase. For that reason, altricial (born in an undeveloped state) young mammals, such as rat puppies, spend 70% of their day in REM sleep, which is proportionally more than, for example, the 20% of bovine calves (Hänninen *et al.*, 2007). Young animals also have a greater need than older animals for energy retention, which can be achieved through more sleep.

3.4 Regulation of Food Intake and Eating

In order for an individual to acquire the appropriate amount and quality of nutrients, the behaviour needs to be optimized and regulated accordingly. Important nutrients include, in addition to energy, such compounds as essential amino acids, vitamins and minerals and, depending on the digestive system and needs of an animal, several behavioural patterns have evolved to make sure those nutritional needs are met.

Finding and ingesting food is a prerequisite for staying alive and thus evolution has resulted in very complex systems to make sure the individual is very motivated to eat. Energy intake is dependent on several internal and external factors, such as the reproductive state of the animal, the amount of exercise and the environmental temperature. Food intake is mainly governed by the hypothalamus, but other mechanisms are also of importance in regulating the food intake. These mechanisms include the level of blood sugar (glucose), the level of filling of the stomach, endocrinological control and cognitive aspects (Schmidt-Nielsen, 1997). As separate mechanisms these cannot regulate the food intake efficiently, but the coordination of several systems is needed for this complex process. For example, ingesting a large bulk of low-energy feed will quickly fill up the stomach, but will not satisfy the energy needs of the animal, which will then compensate by eating an even larger amount of the food. In contrast, a diet with a very high energy content can result in the animal not feeling satisfied due to the lack of filling of the stomach, even though the energy intake might have been appropriate. In times of starvation risk, hypothalamic functions as well as activity in several other parts of the brain will cause food to be an extremely pleasurable reward and thus render the motivation to feed a top priority.

Hormonal control of feeding behaviour

A number of hormones are involved in the regulation of hunger, satiation and feed intake, including hormones such as cholecystokinin (CCK), leptin, somatostatin, insulin and ghrelin (Geary, 2004). CCK inhibits eating and is secreted during meals from the small intestine. The rise in CCK acts in limiting eating at a certain meal, while it does not appear to influence total food consumption. If CCK is administered, meal size is reduced while inter-meal intervals are unaffected or even reduced to compensate for the smaller intake per meal.

Also, injections of leptin can reduce the meal size, but do not appear to affect the inter-meal interval. Leptin is therefore probably more important for determining the body weight than CCK. Leptin is mainly secreted from adipose tissue and the level of leptin is higher in obese individuals. Interestingly, leptin follows a diurnal rhythm and the secretion is not affected by food intake or meal timing. Receptors are found mainly in the hypothalamus, suggesting a role in the homeostatic system, but also in many other parts of the brain. A total lack of leptin will cause immediate overeating and severe obesity. In situations where preferred foods are freely available, individuals can also develop leptin resistance, resulting in obesity. The evolutionary reason for this could be that, in situations of high food availability (e.g. summer), it is of benefit for the animal to consume as much as possible, in order to survive the following period of lower food availability (e.g. winter). Leptin resistance is also seen in animals preparing for hibernation, thus allowing them to eat more than their current nutritional needs call for. Leptin resistance has been found in both obese humans and pet animals, such as dogs and cats, suggesting either that the homeostatic function of leptin is disrupted for some reason in these individuals or that its function simply is not to protect from obesity but to make sure animals of low energy status are highly motivated to eat.

Ghrelin is a hormone produced mainly by endocrine cells in the stomach and it is the hormone with the strongest known positive effect on feeding behaviour. Ghrelin stimulates eating by increasing the feeling of hunger and the ingested meal size. It appears to signal hunger, as its levels increase before meals and decrease after eating. In addition, ghrelin is also affected by the level of adipose tissue – obese individuals have lower levels of ghrelin. Ghrelin receptors in the hypothalamus mediate its effect on eating behaviour, but receptors also exist peripherally, such as in the abdomen.

Insulin is an important anabolic hormone, signalling that nutrients should be stored by the body; it is secreted from the pancreas in response to an increase in blood glucose levels. Thus insulin levels, following an increase in blood glucose, rapidly increase after the ingestion of food.

In addition to the mainly short-term regulation of feeding behaviour discussed above, feeding behaviour is also hormonally regulated on an annual level. For example, melatonin has a marked effect on metabolism and hunger in certain species, such as in the reindeer, where there is a large annual variation in feed availability.

A. Valros and L. Hänninen

Feeding and decision making

Not only changes in the homeostatic balance, such as those induced by food or water restriction, affect feeding behaviour. Many of the decisions related to feeding are related to more complex brain functions, involving such things as memory and the reward system. Throughout evolution finding and consuming food has been of utmost importance, and thus a large portion of the nervous system is involved in this process (Berthoud, 2007). For example, parts of the amygdala appear to be important for learning cues related to food, while the orbitofrontal cortex of the brain is important for memorizing these. This enables an animal to learn which foods are palatable, beneficial or harmful, as well as to memorize where preferred food sources can be found. In humans, simply thinking about food, and in animals giving a cue related to food, causes physiological reactions, such as saliva and insulin secretion (Berthoud, 2007).

The dopaminergic system is involved in the process of wanting foods an animal has learnt to like. When eating a preferred food, dopaminergic neuron activity causes a pleasurable feeling, a reward that later will cause the animal to remember this food as preferred. The cognitive processes described here can cause an animal to want a food even though its metabolic needs do not indicate the need for energy. Therefore, in an environment where food is not a limited resource, animals, as well as humans, easily become obese due to strong evolutionary benefits from a strong motivation to feed.

In addition to meeting the energetic needs of an animal, the feeding behaviour needs to make sure the animal's needs for specific nutrients are met, a process called selective satiation (Berthoud, 2007). It has been shown that certain neurons that become active when ingesting food can distinguish different energy sources from each other. For example, the predictive value of a preferred food is influenced by what foods have been ingested recently: after a high-glucose meal there is less neural activity when presented with further sugar-rich foods, while other foods might still cause neural activity of similar magnitude.

3.5 Reproductive and Maternal Behaviour

Hormonal regulation of reproductive behaviour

Reproductive behaviour is largely controlled by endocrinological sequences, although these processes are also influenced by external factors such as season, food availability and social factors. In most animals, timing of reproduction is set at a certain time of the year, when the survival possibilities for the offspring are maximized. Typically, offspring are born in the spring or early summer, when food resources are usually available. The seasonality of the reproductive cycle is mainly regulated by light–dark cycles, something that is utilized for example in the egg industry: under natural lighting conditions laying hens decrease egg-laying during the winter but, by increasing the daylight with artificial lighting, egg production can be kept more or less stable throughout the year. Also, by using different light rhythms during rearing, the lifetime production capacity of modern laying hens can be further refined and optimized.

In other production animals the seasonality of reproduction has negative impacts on optimizing the production capacity: in sheep, all-year-round breeding is very challenging and in pigs remnants of seasonality can be seen as lowered piglet production during the autumn period, even though the domestic pig is no longer a purely seasonal breeder. Also, in the boar, a seasonal variation can be seen as higher levels of testosterone, increased quantity and quality of sperm and a related higher level of sexual activity in the autumn, which is the main breeding season of the pig. It appears that it is the length of the light period, not the intensity, that is important for governing this system.

The effect of light is most probably transmitted via melatonin to the hypothalamus, which, as stated earlier in this chapter, is the main regulator of the reproductive functions of animals. Two of the main hormones involved in the regulation of reproductive cycles and behaviour are the gonadotropins, i.e. LH and FSH. The main target organs of the gonadotropins include the ovaries in females and the testes in males. In females, the release of gonadotropins mainly affects the release of oestrogen, influencing the development of secondary female sex characteristics and sexual behaviour. Infusions of oestrogen into the brain of castrated cats can, for example, induce behaviour indicative of willingness to mate. In the male, the most potent androgen is testosterone, which affects secondary sex characteristics, such as plumage and facial hair growth, as well as sexual behaviour. Testosterone affects the level of aggressiveness in the male, as well as the level of sexual activity.

Hormonal regulation of maternal behaviour

The maternal behaviour of animals is largely under endocrinological control. Behaviour during the prepartum period, such as nest building in the sow, during parturition itself and during the postpartum period (including both nursing behaviour and behaviour related to the mother–young bond), has been shown clearly to have a hormonal background. This does not, of course, mean that environmental factors would be of no importance, but does indicate that behaviours related to maternal abilities are highly motivated in the dam. In addition, continuous feedback from the offspring is needed for the continuance of maternal behaviour. For example, as a result of reduced udder manipulation, milk production and related hormone secretion will decrease prematurely.

During pregnancy the level of the steroid hormone progesterone is high, acting to maintain the pregnancy. Progesterone originates from the corpora lutea in the ovary of a pregnant female. Prior to parturition, progesterone levels drop, thus allowing milk production to begin and metabolic functions to favour milk production instead of storage of nutrients in adipose tissue. If the level of progesterone does not decrease appropriately, this can have adverse effects on milk production and thus on the growth of the nursed offspring.

Before parturition, levels of prolactin start to increase. Prolactin is one of the main hormones involved in initiating and maintaining milk production, as well as in influencing the maternal and nursing behaviour of the dam. Prolactin also affects lactational metabolism by increasing the amount of insulin receptors in the mammary gland. High levels of prolactin are maintained by stimulus from the offspring and, if this is reduced, prolactin levels will also decrease.

In the pig, one of the most marked effects of prepartum prolactin is the effect on the sow's nest-building behaviour (see Figure 3.5). A rise in prolactin a few days before farrowing causes the initiation of nest-building behaviour. This

A. Valros and L. Hänninen

Fig. 3.5. The domestic sow, like its ancestor the wild boar, builds a nest prior to farrowing in order to provide shelter for the piglets. This nest-building behaviour is largely regulated by prepartum endocrinological changes (image courtesy of Heli Castrèn).

pre-farrowing increase in prolactin is related to an increase in prostaglandin (PGF2α), which also increases significantly prior to farrowing. Different aspects of the nest-building behaviour of the sow have also been shown to be correlated with concentrations of other hormones, such as progesterone, somatostatin and oxytocin. In addition, the sow alters her nest-building behaviour depending on environmental cues, such as temperature. The mechanism behind the termination of nest building is less well known, and probably involves complex hormonal and environmental interactions. A high level of oxytocin just before parturition has been shown to correlate with an early cessation of nest building, but the mechanism for this is still unknown.

Oxytocin is secreted from the hypophysis and is important for several different aspects of maternal behaviour in mammals. For example, grooming of pups in the rat is induced by oxytocin. Oxytocin is needed for the initiation of lactation and for milk ejection during separate nursings. Stimulation by the offspring, such as udder manipulation, stimulates oxytocin release, which then triggers milk 'let-down'. In addition to this all-or-nothing effect on lactation, oxytocin influences metabolism during lactation, ensuring that resources are available for milk production. In the pig, a high basal oxytocin level has been linked to a high catabolic level (negative energy balance) of the sow, and a correlated high weight gain in the piglets (Valros, 2003). In addition, oxytocin has an important role in the parturition process, by stimulating uterine contractions. In general, the expulsion of each fetus is related to a peak in oxytocin, but it also appears that in litter-bearing species, the birth of the first fetuses stimulates the parturition process further. In pigs and rats, a high oxytocin level has also been linked to a shorter duration of parturition (Algers and Uvnäs-Moberg, 2007).

Hormones related to feed intake and metabolism have also been shown to be important in the regulation of lactation and nursing behaviour. As insulin promotes the utilization of nutrients for peripheral body tissues, its secretion is decreased during stages where catabolism can be seen as an advantage, such as during lactation, where nutrients should be directed mainly towards milk production. For example in the sow, insulin pre-feeding levels decrease with proceeding lactation, and a correlation between sow nursing behaviour and insulin levels has been reported. Sows with high levels of pre-feeding insulin show more behaviour related to avoiding udder massage by their piglets. It is possible that a high level of nursing stimulus increases the number of insulin receptors in the mammary gland, thus decreasing the level of circulating insulin, but the possibility cannot be excluded that this high insulin level, as such, influences the nursing motivation of the sow, possibly via interaction with prolactin. Whatever the cause and the effects, it appears important for the sow to be in a highly catabolic state during lactation, as this appears to be a prerequisite for efficient piglet growth and high piglet survival rates (Valros, 2003).

Recent studies also suggest that the level of leptin influences the nursing behaviour of sows and piglets, as the leptin level in sows' milk is negatively correlated to nursing frequency and piglet weight gain (Oliviero et al., 2006). Hormones such as vasoactive intestinal polypeptide (VIP), gastrin, somatostatin and glucagon are also influenced by suckling. All of this causes the optimization of nutrient utilization for milk production by affecting gastrointestinal functions and allocation of nutrients.

3.6 Stress-related Behaviour

When an individual is subjected to a stressor, the blood concentration of glucocorticoids increases as a result of activation of a specific endocrinological pathway, namely the hypothalamic–pituitary–adrenocortical (HPA) axis. When exposed to a stressor, C-RH from the hypothalamus stimulates pituitary ACTH secretion, which leads to an increased glucocorticoid release from the adrenal cortex. In addition, the sympathetic nervous system increases an individual's performance in demanding situations by increasing the secretion of catecholamines, such as adrenalin and noradrenalin, from the adrenal medulla. Noradrenalin is also secreted directly from the sympathetic nervous system, being the terminal transmitter substance of this system. These reactions will cause the so-called fight-or-flight response, preparing the individual to face physical or physiological challenges. During the acute fight-or-flight response, the blood pressure increases and blood sugar concentration rises. Heart frequency increases and blood oxygen is efficiently transported, especially to the skeletal muscles. Individuals are bright and alert, preparing themselves either to fight, flee or freeze. However, the secretion of glucocorticoids in itself is not a sign of a negative experience, as it is also stimulated by physical exercise, feeding or mating. It is also worth noting that, if the stressful situation is prolonged, it disturbs the circadian variation of glucocorticoids in the blood.

For more information on stress-related mechanisms, see Chapter 7.

A. Valros and L. Hänninen

References

Algers, B. and Uvnäs-Moberg, K. (2007) Maternal behaviour in the pig. *Hormones and Behaviour* 52, 78–85.

Berthoud, H.-R. (2007) Interactions between the 'cognitive' and 'metabolic' brain in the control of food intake. *Physiology and Behaviour* 91, 486–498.

Buijs, R.M., van Eden, C.G., Goncharuk, V.D. and Kalsbeck, A. (2003) Circadian and seasonal rhythms: the biological tunes the organs of the body: timing by hormones and autonomic nervous system. *Journal of Endocrinology* 177, 17–26.

Cardinal, R.N., Parkinson, J.A., Hall, J. and Everitt, B.J. (2002) Emotion and motivation: the role of the amygdala, ventral striatum and prefrontal cortex. *Neuroscience and Biobehavioural Reviews* 26, 321–352.

Dawkins, M.S. (2006) Through animal eyes: what behaviour tells us. *Applied Animal Behaviour Science* 100, 4–10.

Eckert, R. (1988) *Animal Physiology. Mechanisms and Adaptation*, 3rd edn. W.H. Freeman and Company, New York.

Geary, N. (2004) Endocrine controls of eating: CCK, leptin and Ghrelin. *Physiology and Behaviour* 81, 719–733.

Hänninen, H., Hepola, H., Raussi, S. and Saloniemi, H. (2007) Effect of colostrum feeding method and presence of dam on the sleep, rest and sucking behaviour of newborn calves. *Applied Animal Behaviour Science* 112, 213–222.

Hänninen, L. (2007) Sleep and rest in calves – relationship to welfare, housing and hormonal activity. PhD thesis, Helsinki University, Finland.

Janczak, A.M., Braastad, B.O. and Bakken, M. (2006) Behavioural effects of embryonic exposure to corticosterone in chickens. *Applied Animal Behaviour Science* 96, 69–82.

Janczak, A.M., Torjesen, P., Palme, R. and Bakken, M. (2007) Effects of stress in hens on the behaviour of their offspring. *Applied Animal Behaviour Science* 107, 66–77.

Oliviero, C., Heinonen, M., Valros, A. and Peltoniemi, O.A.T. (2006) High concentration of leptin in the sow's milk during lactation seems to reduce the frequency of feeding and the weight gain of piglets. *Reproduction in Domestic Animals* 41, 367.

Opris, I. and Bruce, C. (2005) Neural circuit of judgement and decision mechanisms. *Brain Research Reviews* 48, 509–526.

Ressler, N. (2004) Rewards and punishments, goal-directed behaviour and consciousness. *Neuroscience and Biobehavioural Reviews* 28, 27–39.

Ruckebush, Y. (1972) The relevance of drowsiness in the circadian cycle of farm animals. *Animal Behaviour* 20, 637–643.

Schmidt-Nielsen, K. (1997) *Animal Physiology. Adaptation and Environment*, 5th edn. Cambridge University Press, Cambridge, UK.

Siegel, J.M. (2005) Clues to the functions of mammalian sleep. *Nature* 437, 1264–1271.

Valros, A. (2003) Behaviour and physiology of lactating sows – associations with piglet performance and sow postweaning reproductive success. PhD thesis, Helsinki University, Finland.

Zepelin, H., Siegel, J. and Tobler, I. (2005) Mammalian sleep. In: Kryger, M.E., Roth, T. and Dement, W.C. (eds) *Principles and Practice of Sleep Medicine*. Elsevier–Saunders, Philadelphia, Pennsylvania, pp. 91–100.

4 Motivation and the Organization of Behaviour

G. MASON AND M. BATESON

4.1 What is Motivation and Why is it Important?

Motivation as a 'causal' explanation for animal behaviour

Motivational states include thirst, hunger, fear and the urges to migrate, mate, nest-build and dust bathe. These motivations are internal states which vary in magnitude from one moment to the next, and that help determine which stimuli animals will react to and which they will not; which goals they seek out and which they do not; and the effort or intensity with which they will perform a given behaviour. Motivations are examples of causal explanations for behaviour (see Chapter 1), meaning that they are a proximate, mechanistic explanation for why an animal is currently performing a particular behaviour pattern.

Not all changes in behaviour are due to changes in motivation. We do not invoke motivation to explain behaviours that are always elicited in the same way by the same stimuli (e.g. reflexes like blinking, or rapid limb withdrawal from painful stimuli); changes in the cues that animals respond to as a result of learning (see Chapter 5); or many of the behavioural changes that result from developmental processes (e.g. maturation) or illness/injury (e.g. ceasing to eat because of a gum abscess). Instead, the properties of animal behaviour that motivation helps explain can be summarized as follows. First, changes in motivational state help explain the decisions made by individual animals when faced with choices about what activity to do next. Thus, if you watch an animal for any period you will notice that it engages in different behaviour patterns, apparently spontaneously switching between these at intervals. For example, in the middle of the day, domestic chickens switch from foraging to dust bathing; while a domestic cat during the course of its day will switch from feeding to drinking, from hunting to resting, and so on. Changes in motivational state also help explain why sometimes external stimuli act as powerful triggers for behaviour, but at other times are effectively ignored. For instance, if a hen is deprived of a suitable substrate in which to dust bathe, she will dust bathe vigorously when she is finally allowed access to, say, wood shavings or peat. However, once she has finished doing this, for a while afterwards that very same substrate will lose its ability to stimulate her to perform any further dust bathing. Finally, when watching our deprived hen indulge in dust bathing, we might well notice an increase in this activity's duration and intensity. Thus, along with the increased probability of occurrence of a specific behaviour pattern, we typically see other changes as well, such as increased rates of performance and, if obstacles are placed in an animal's way, increased efforts made to perform the behaviour.

Motivational states as intervening variables

From the section above, and perhaps from our own personal experiences, we can see that strong motivations alter behaviour in many different ways. For example, if we look at drinking behaviour we notice that, as the motivational state of thirst increases, the probability that drinking will occur also increases; but also the range of liquids found acceptable increases (for instance, rats will tolerate quinine in water if they are thirsty – a bitter substance they otherwise avoid), the latency to start drinking decreases (i.e. it is initiated more rapidly when liquid is presented) and the rate of drinking and the amount consumed both increase. If the animal has to pay a cost to drink, it will also be more willing to do so, and this might involve learning novel responses: thus, just as hot and thirsty humans will pay inflated prices for cold sodas and bottled water, so will thirsty rats press a lever multiple times to gain water. As we discuss later, one reason for animals being prepared to perform effortful responses is that being unable to perform highly motivated behaviours is frustrating and accompanied by negative emotions, whereas performing them is accompanied by positive ones. Thus, serious thirst is subjectively unpleasant, while water also tastes wonderful when we are thirsty (see, e.g., Rolls, 2005).

The suite of behavioural and emotional changes described above is, in turn, induced by a suite of diverse events or 'inputs' originating from physiological signals from the body (e.g. hormonal changes, signals of homeostatic need) and sensory inputs from the external world. For instance, the motivation to drink can be elicited by just having eaten, by a dry mouth, by hypernatraemia (high blood sodium levels), by the sight of another animal drinking, by the smell of a very palatable drink or by the sight or sound of other external cues that, through learning (see Chapter 5), have come to be associated with drinking.

If we were to draw all these various possible causal links outlined above, we would come up with a very complicated diagram (see Fig. 4.1a). A far simpler, more economical way of summarizing the many observed relationships is instead to postulate a single 'intervening variable' – in this case something called 'thirst' – that is influenced by *all* the many independent variables (dry mouth, external cues and so on) and in turn influences *all* the dependent variables (amount consumed, perceived pleasantness of

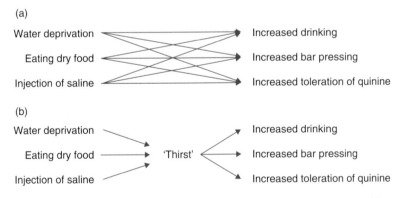

Fig. 4.1. Two alternative models for the explanation of behaviour. Three causal factors on the left (water deprivation, ingestion of dry food and injection of saline) lead to three types of response on the right (increased drinking, increased bar pressing for water and increased toleration of quinine in available drinking water). Miller noted that explanations could involve postulating nine stimulus–response links (a) or a smaller number of links if an intervening variable ('thirst') is included (b) (after Miller, 1956, in Barnard, 2004).

drinking and so on; see Fig. 4.1b). Motivational states are thus best conceptualized as intervening variables that provide an integrative link between various functionally related inputs and behavioural outputs. This general concept underlies the more complex models of motivation described in this chapter. But, first, how and where does this 'integrative link' occur?

Where and how do motivational states occur?

Since it is the brain that drives and controls behaviour, it is a truism that motivational states arise in the brain. Indeed, a 'motivational state' can be defined as the combined physiological and perceptual state of an animal, as represented in its brain. Beyond acknowledging this, however, ethologists have traditionally treated the animal as a 'black box', meaning that they are not interested in the details of the mechanisms inside it. Under this approach, the focus is more on understanding the 'rules' that translate the current internal state of the animal and the external stimuli to which it is exposed in observed behaviour. This 'black box' approach can be useful because it can help us to model, quite simply, how it is possible for the same cues to elicit different responses at different times and in different individuals. To illustrate this, if we understand how dust bathing is affected by the oiliness of feathers and the length of deprivation from a bathing substrate, we can then understand why some birds might have higher motivation to dust bathe than others when presented with an appropriate substrate. Likewise, if we can record how animals weigh up the costs and benefits of performing particular behaviours (e.g. gaining water to drink versus expending time and energy lever pressing for it), we can model and predict many aspects of observed behavioural decision making. Knowing the exact internal physiological mechanisms by which these processes occur would not necessarily yield information with any more explanatory power.

However, in other instances, understanding the precise physiological mechanisms giving rise to motivational states *can* be very useful: opening the black box to dissect its endocrine and neural mechanisms may give us unique insights. In his work on periparturient sows, for example, Gilbert has shown that these animals' nesting motivations are responses elicited within the brain by prostaglandins that are released as part of physiological preparation for birth. By showing exactly how motivations to perform this specific natural behaviour arise, this mechanistic research reveals how unlikely it is that we will be able to reduce this motivation by any intervention other than by allowing sows actually to build nests (see, e.g., Boulton *et al.*, 1997).

Neurobiologists have also recently become interested in the mechanisms underlying motivational states. This exciting area of research is fuelled, not by a desire to understand animal behaviour, but instead by the need to understand problematic 'over-motivated' behaviours in humans: compulsions and addictions such as overeating, gambling and drug taking. It is beyond the scope of this chapter to review the neurobiology of motivation. However, in brief, forebrain regions such as the orbitofrontal cortex and the basal ganglia's nucleus accumbens have been found to control the effort and intensity with which motivationally relevant stimuli are sought out and responded to, and have also been implicated in addiction and abnormal compulsive behaviours with an emotional component. Berridge (2004) and Rolls (2005) provide good further reading for those wishing more detail. Applied ethologists are now starting to find this mechanistic information useful for investigating abnormal behaviour

G. Mason and M. Bateson

in captive animals (see below and Chapter 7). Some parallel work by psychologists on both animal and human subjects has also helped clarify the links between motivation, emotion and learning – something with important animal welfare implications, as we discuss below.

Why it is important to understand domestic animals' motivations

Understanding animal welfare and the origins of abnormal behaviour are two of the main reasons why it is important to investigate domestic animals' motivations. Motivational research helps us identify – and hopefully alleviate – the sources of some important frustration-related welfare problems; it also helps reveal why abnormal activities like stereotypic behaviours are so common in captive animals. However, even if we are not interested in welfare, understanding motivation can still be important practically, because it can aid the effective management of both learnt and non-learnt aspects of behaviour. For example, if we appreciate how both the nature of internal states and the quality of external cues affect motivation, we can use this to design training regimes where the rewards offered are most likely to cause animals to learn what we want them to learn. We also can use this understanding to manipulate instinctive aspects of behaviour such as mating and maternal care. We expand on all four of these topics in Section 4.4, below.

4.2 Motivational States and the Organization of Behaviour

The interaction of internal and external stimuli

Motivational states can be induced by internal stimuli (signals from the body), external stimuli (cues in the outside world) and the interactions between them (i.e. how they combine to influence the animal).

Perhaps the most obvious type of internal stimulus to affect motivation is one that reflects a disruption of homeostasis, leading to motivations to perform behaviours that will restore an animal's physiological equilibrium. Thus cues from the body signalling dehydration induce strong motivation to drink, while cues from the body signalling low energy induce motivation to eat. Other common examples of internal states important for motivation include reproductive hormones. For example, in terms of performing operant responses like lever pressing, or overcoming aversive barriers like electrified floor-grids, female rhesus monkeys will work harder to reach a male when they are in oestrus than when they are not. Female marmosets will also lever press for the sight and sounds of an infant, but only when they are heavily pregnant or lactating – an effect mediated by oestrogen. The role of hormones in the nest-building motivation of sows has already been touched upon, and is developed further in Chapter 3.

External stimuli that affect motivation include signals of time of day or seasons of the year (the latter often exerting their effects via endocrine changes), and cues that signal a motivationally relevant resource. These are sometimes known as 'eliciting stimuli'. We are all familiar with the way the smell of a delicious meal can make us feel ravenously hungry. Eliciting cues can likewise have powerful effects on animals' motivations, in behavioural systems as diverse as feeding and mating. They are particularly important in triggering escape motivations (elicited by predator cues), aggressive motivations (elicited by cues from rival conspecifics) and, in cats and other carnivores, predatory responses like pouncing and biting (elicited by the movements, squeaks and struggles of prey). Sometimes animals will habituate to one type of cue, only to find a new one powerfully motivating: thus the presentation of new sexual partners, or of new types or flavours of food, may boost or renew mating or feeding motivations (a phenomenon known as the Coolidge effect, after the American president of the same name). Note that learnt cues (e.g. the sight of a favourite restaurant to humans; or sound of the fridge door opening to a cat) can affect motivational states in this way too.

Most motivated behaviours are affected by a combination of external cues and internal states. Classic work by Baerends *et al.* (1955, described in McFarland, 1985) on courtship behaviour in male guppies demonstrates how the size of the female fish (an external stimulus that relates to her attractiveness and fertility) and the colour patterns on a male (an external indication of the male's internal sexual state, especially testosterone levels) combine to affect his behaviour. Male guppy courtship proceeds through a series of stages from posturing, via intentional sigmoid displays to full sigmoid displays. Figure 4.2 shows that the attractiveness (i.e. size) of the female required to elicit a given display declines as the internal mating motivation of the male increases. The principles revealed in this experiment apply to many other behaviour systems. For example, food-deprived animals need only poor-quality cues to elicit feeding, while sated animals will respond only to eliciting stimuli that signal highly preferred, palatable or novel food items.

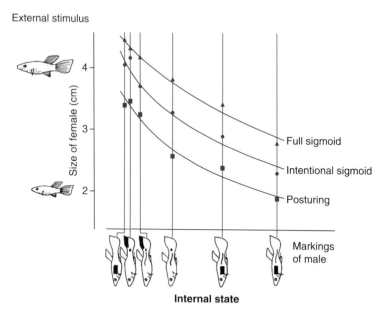

Fig. 4.2. The size of a female guppy needed to elicit each of three different displays (posturing, intentional sigmoid and full sigmoid) from male guppies as a function of the markings on the male. The size of the female is an external stimulus and the markings on the male are an indication of his internal state. The curved lines plot 'motivational isoclines' – lines that connect points of equal likelihood that a male will perform a given display (modified from Baerends *et al.*, 1955, in McFarland, 1985).

G. Mason and M. Bateson

The shape of the motivational isoclines in Fig. 4.2 may tempt us into thinking that female attractiveness and male sexual motivation can be multiplied together to predict courtship behaviour in male guppies. Could this be a general principle for how external and internal factors interact in other motivational systems and other species? Sadly, the biology of motivation is not this simple. In the case of guppy courtship we need to be very careful in interpreting the shape of the curves, because the measurement of the males' markings was based on a rank order as opposed to a true interval scale.

More generally, the relative role of internal and external factors and the precise way in which they interact to produce behaviour appear to vary extensively between different motivational systems. Most seem to require both, but there is a spectrum, with some motivated behaviours affected predominantly by internal stimuli and others primarily by external stimuli. Examples of the former include the way a rise in testosterone levels is sufficient to start singing behaviour in a male songbird such as a starling; the nest building of sows is another case in point. In contrast are those motivated behaviours affected predominantly by external stimuli. One example here would be explorative behaviour in mice, which is not seen at all when an animal is in its home cage, but is elicited by placing it in a novel environment; other examples, as already mentioned, include aggression and anti-predator behaviour. A recent experiment with caged mink clearly demonstrates how external factors are more important for some motivational systems than for others. Warburton and Mason (2003) compared individual minks' motivation to reach four different resources in parallel (food, swimming water, toys and social contact). Animals had to push one of four weighted doors (depending on which resource they were motivated to access), and then travel along a tunnel 7.5 m long to reach the resource. This tunnel either doubled around to reach resources that were visible from the animal's home cage, and were thus visible when it pushed the weighted door ('cues treatment'); or it went behind an opaque screen to where the various resources were placed out of sight from the home cage ('no cues treatment'; see Fig. 4.3a). Whether or not cues were present when the mink chose to push the door affected their motivation to visit some of the resources. When their cues were not detectable from the home cage, mink visited toys and a conspecific less often than they did food, but when cues were detectable they visited them as often. In contrast there was no effect of cues on visits to food or swimming water (see Fig. 4.3b). Furthermore, the minks' 'consumer surplus' (a measure of value or motivational importance) for toys, but no other resource, was significantly lower when the toys' cues were screened compared with when they were visible at the choice point. Thus the effect of external stimuli varies depending on which motivation you are dealing with.

The odd behavioural responses that we see in animals in some artificial situations can be very revealing about motivational mechanisms underlying behaviour. Animals deprived of the opportunity to perform behaviour patterns will sometimes show so-called 'vacuum activities', behaviour patterns occurring in the absence of the normal external eliciting stimuli. For example, hens housed in battery cages with wire floors will still go through the actions of dust bathing despite the absence of any suitable substrate. At the other end of the spectrum, external stimuli can sometimes alone be so powerful that they can trigger inappropriate behaviour. A herring gull will retrieve eggs that have rolled out of its nest but, when given a choice of two eggs, one of natural size and one 50% larger, it will preferentially retrieve the larger egg. The larger egg is a good example of what ethologists call a 'supernormal stimulus', because it is more effective at eliciting behaviour than the natural stimulus. Supernormal stimuli from conspecifics

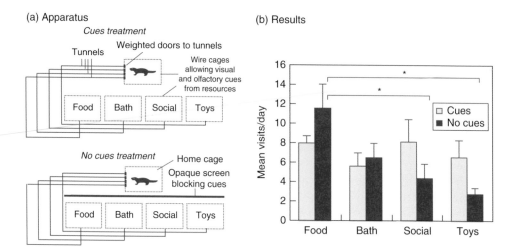

Fig. 4.3. An experiment to explore the effect of the availability of external cues on minks' use of four resources (food, swimming bath, access to social contact and toys). (a) The apparatus used in the two treatments of the experiment; in the 'no cues' treatment the screen blocks cues from the resources at the time the mink decides whether to work to access a resource; (b) the mean visits per day (one simple measure of motivation) made by the mink to each of the resources. Asterisks indicate significant differences (after Warburton and Mason, 2003, who also show further data on other indices of motivation).

are likely to be very common in captive environments due to the high densities at which animals are housed, thus amplifying the normal cues provided by other animals in the environment.

Regulatory processes in the motivational control of behaviour: negative feedback processes, hysteresis, positive feedback and switching costs

Behaviour usually occurs in discrete bouts. For example, animals typically eat in 'meals', as opposed to nibbling constantly (lions can eat 30% of their own body-weight in one sitting!) and, in the hen, dust bathing typically occurs in bouts of about 20 min that are performed every 2 days. Therefore, some motivational mechanism is needed to explain why a bout of behaviour continues as it does, and then ends. Negative feedback is the process whereby execution of a behaviour pattern reduces the motivation to perform it, and is important in limiting the length of bouts of many different behaviours. Negative feedback may come from the direct consequences of the behaviour. For example, the absorption of water into the body post-drinking causes a cascade of physiological consequences that reduce motivation to drink. Negative feedback may also come from the performance of a behaviour pattern *per se*. For example, the satiating effect of water injected into the body intravenously is not as great as water drunk by mouth. Similarly, mutant featherless hens will still go through the actions of dust bathing despite the fact that the behaviour can be having no effect on the condition of their (non-existent) plumage. This evidence supports the

G. Mason and M. Bateson

hypothesis that dust bathing has rewarding consequences for chickens above and beyond its direct hypothesized function in feather maintenance.

Whereas negative feedback is important in stopping animals from performing a behaviour pattern indefinitely, mechanisms are also necessary to make sure that animals persist at a behaviour for long enough to achieve its functional goal. A simple, competition model of motivation whereby an animal always immediately switches to the behaviour with the highest motivation suffers from the problem that it will leave animals constantly 'dithering' between, for example, tiny bites of food and tiny sips of water (see Fig. 4.4). A range of solutions has been suggested to deal with this problem, since real animals do not typically behave like this.

Delayed negative feedback (or 'hysteresis'), whereby it takes a period of time for negative feedback, to reduce motivation, is one feature of motivational control that arrests dithering and gives bouts some stability. Furthermore, this is often aided by positive feedback, which, although similar in effect, differs in that motivation actually *increases* after the start of a bout. One good example comes from careful studies of eating in mice by Wiepkema (1971). He analysed the 20–30 mini-bouts of eating that comprise each single meal a mouse eats. When mice are food deprived for 24 h, mini-bouts increase in length (from about 8 to 13 s) and the gaps between them decrease (from about 14 to 9 s), illustrating an effect of deprivation. However, *within* each meal, and particularly at the start, mini-bouts initially also increased in length and occurred more closely together (thus, for example, the fifth mini-bout of a meal is longer than the first), nicely illustrating the effects of positive feedback. Later on in

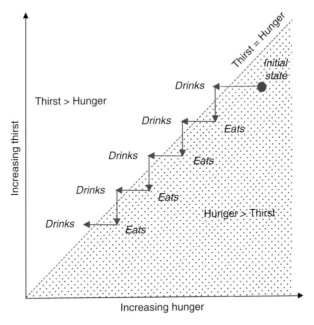

Fig. 4.4. Dithering between eating and drinking behaviour predicted by a simple competition theory of motivation. The animal is assumed to reduce its initially higher motivation, in this case hunger, until thirst is greater than hunger (i.e. the dotted line has been crossed), at which point it starts drinking. This model does not describe well what animals typically do in realistic situations (for explanations, see text) (modified from McFarland, 1985).

the meal, the mini-bouts then decrease in duration again, illustrating the negative feedback that eventually terminates the meal.

Another factor affecting the length of behavioural bouts is the cost of switching to a different behaviour. Animals stick longer at a single type of behaviour (A) if it is costly in terms of time, energy or potential danger to switch to another (B), perhaps because the motivation for B has to build up to a higher level before the animal will be prepared to pay high costs to switch. A classic example of this phenomenon is provided by McFarland's (1971) study (described in McFarland, 1985) of switching between eating and drinking in doves. He showed that, if the birds had to negotiate a barrier to switch between these two behaviours, then they switched less often – and so had longer bouts – when the barrier was large than when it was small. It appears, therefore, that animals take into account the cost of switching activities when decision making.

If the explanation for the delay in switching is that motivation has to build to a higher level before the animal is prepared to pay the switching cost, then this higher motivation should be reflected in the behaviour of the animal when it finally makes the switch. Data supporting this prediction are provided by an experiment in which mink were offered an array of different enrichments to interact with, but the costs of reaching each were increased by weighting the doors that the mink had to push to reach them. Increased costs resulted in two main changes in behaviour: first, as with McFarland's doves, the animals switched between enrichments less often, using each enrichment for longer in each bout; second, when they eventually paid the cost to use a new enrichment, they had a *shorter* latency to interact with it after pushing open the door. This latter result illustrates that, when the costs of switching were higher, a higher motivation had built up before the animal chose to pay the cost of switching (Cooper and Mason, 2000).

Behavioural sequencing: appetitive and consummatory behaviour and feed-forward processes

Bouts of motivated behaviour rarely comprise the exact same motor sequence repeated again and again, but usually have some kind of internal structure to them. For example, a bout of dust bathing in a chicken starts with the bird scratching and bill-raking at the substrate. Next it erects its feathers and squats in the substrate and performs a sequence of vigorous dust bathing behaviours. This phase is followed by the bird flattening its feathers and lying on its side. Finally it stands up and shakes the substrate from its plumage before switching to another activity. All of these phases within the sequence appear to be motivated by a single state.

A distinction is often made between 'appetitive' and 'consummatory' phases within a behavioural sequence. The appetitive phase comes first, and comprises active, flexible, searching behaviours; appetitive behaviour can even include novel responses that the animal learns, e.g. lever pressing. Appetitive behaviour may appear quite similar even when motivated by quite different states. For example, the appetitive phases of food search, mate search and nesting site search are all superficially quite similar both within and between species. Appetitive behaviour is followed by consummatory behaviour, which is more stereotyped, unlearnt, species-typical and motivation-typical. Examples include the specific movements used in foraging (e.g. cows and giraffes tongue-twirl, sheep and goats bite, pigeons peck); drinking

G. Mason and M. Bateson

(e.g. cats and dogs lap, pigs and humans suck, starlings dip-and-tilt); and mating (e.g. lordosis by female quadripeds, thrusting by males). The term 'fixed action pattern' is often used for the stereotyped, species-typical forms involved. Analysing fixed action patterns can be a useful way of identifying the motivational bases of abnormal behaviour, as we will see in Section 4.4.

In addition to the predictability within a sequence of motivated behaviour, transitions between one type of motivated behaviour and another are also often rather predictable. Thus, often one behaviour pattern follows another: drinking follows eating (in, e.g., rats and humans) and may be followed by thermoregulatory behaviour (if drinking reduces core body temperature); grooming follows eating in cats; and preening follows water bathing in birds. Such adaptive sequences of behaviour may be partially the product of 'feed-forward' mechanisms. Traditionally, feeding and drinking behaviour were thought to be controlled by feedback mechanisms that tracked the levels of physiological indices of hunger and thirst (see above). However, eating typically motivates subsequent drinking even *before* the osmotic consequences of food intake have commenced, suggesting that feed-forward mechanisms may be more important in the control of feeding and drinking than previously thought. In at least some instances, such feed-forward effects are the product of past experience and learning (e.g. Ramsay *et al.*, 1996).

4.3 Models of Motivation

In order to conceptualize how internal and external factors are integrated, and how negative feedback, positive feedback and other processes occur, ethologists have modelled motivational states in a number of different ways. As we shall see, different models concentrate on explaining different aspects of the observed behavioural phenomena that we have described above.

The psychohydraulic model

One early model of motivation was that of Lorenz (1950, described in McFarland, 1985). His 'psychohydraulic model' likens behavioural control to a cistern of water that can overflow via a valve, this overflow representing the performance of behaviour (see Fig. 4.5). The cistern steadily fills from an in-pipe, representing increasing levels of internal stimuli when a behaviour is not performed. Overflow via the valve occurs if a combination of the pressure due to the build-up of water in the cistern (representing the build-up of internal stimuli) and weights pulling on the valve (representing the influence of external stimuli) reaches a critical threshold. Flow ceases again once the fluid level in the cistern and hence the internal pressure have decreased.

This model admirably represents the ways that internal and external factors combine to determine motivation; that behaviour patterns occur in predictable sequences; and that several aspects of behaviour change with increased motivational state, including duration, rate and the types of behaviour patterns produced. However, one criticism of this model is that it implies that the motivation always remains high unless behaviour is performed, because overflow via the valve is the only way for the cistern to empty. While this may be appropriate for certain behaviours (e.g. eating and drinking), it clearly is not the case for others: as we have already discussed,

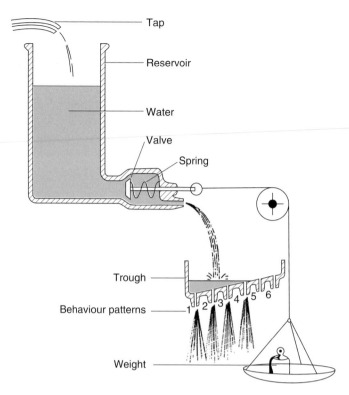

Fig. 4.5. Lorenz's psychohydraulic model of motivation. Behaviour occurs when water exits the reservoir into the trough. The spring holds the valve shut until the valve is opened by a combination of the pressure of water that has built up in the reservoir (i.e. internal state) and the weight in the pan (i.e. external stimuli). The behaviour patterns displayed (1–6) depend on the level of water in the trough, with 'higher' patterns requiring higher levels of water (modified from Lorenz, 1950; in McFarland, 1985).

internal factors seem unimportant for some behaviours. In the absence of rivals, for instance, an animal will not become more and more likely to fight, and likewise, if predators are absent, an animal will not become more likely to hide or make alarm calls. Thus for many behaviours, if the goal of a behaviour exists already, the motivation to perform that behaviour will remain low.

The Sollwert–Istwert model

That same year, von Holst and Mittelstaedt (1950, described in Mason *et al.*, 1997) and colleagues developed an alternative model that did not assume an inevitable build-up of internal causal factors in the absence of behavioural expression. Their 'Sollwert–Istwert' model conceptualizes a motivational system as acting more like a thermostat, causing behaviour that brings an animal's state (its 'Istwert' – literally, 'the way the world is') closer to a desired end point (the 'Sollwert' – 'the way the world should be'). The greater the discrepancy between Sollwert and Istwert, the stronger the motivation to perform the behaviour.

G. Mason and M. Bateson

The important difference between this and the psychohydraulic model is that it emphasizes that motivation may be extremely low if the end point of the behaviour is already in existence. However, a similarity between the two models is that negative feedback is the key means of behavioural control. Since this is too simple to account for how real animals partition their time between different activities (see above), the Sollwert–Istwert model was used as the starting point for more sophisticated representations of the control and sequencing of behaviour, considered next.

Systems models

So-called 'control systems' models of motivation are based on the assumption that we can draw analogies between the mechanisms of living and non-living systems (McFarland, 1971, described in McFarland, 1985). These models incorporate means of control other than negative feedback so that, for example, motivation can increase in anticipation of a discrepancy between Sollwert and Istwert (i.e. a feed-forward process), rather than always reacting to discrepancies after the event. These models represent behaviour using systems diagrams or control theory from engineering (see Fig. 4.6). In contrast to the models above, they can represent behavioural sequences (e.g. appetitive and consummatory phases) and also these often incorporate stages of feed-forward, hysteresis and positive feedback. These models also incorporate something valuable from the psychohydraulic model, i.e. the performance of behaviour *per se* can, at times, be important in reducing motivation. Furthermore, Hughes and Duncan (1988) did a valuable job in refining these models and used them to discuss 'ethological needs' and abnormal behaviour in captive animals, topics we return to in Section 4.4.

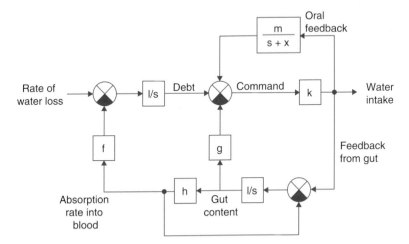

Fig. 4.6. A control systems model of drinking in doves. In this model oral stimulation during drinking provides positive feedback, and gut factors provide negative feedback. The circles represent points in the control system at which different variables are assumed to be summed up, and a red quadrant in one of these circles changes the sign of an input. The letters within the boxes correspond to parameters of the various component mechanisms (modified from McFarland and McFarland, 1968, in McFarland, 1985).

State–space models

The next generation of motivational models focused in more detail on the interactions between behaviour patterns and the rules animals follow when choosing between different activities. State–space models (McFarland and Sibly, 1975; described in Barnard, 2004) represent an animal's motivational state as an n-dimensional vector resulting from the interaction between internal and external factors. This motivational state is the state value of all the causal factors influencing a set of functionally related behaviour patterns. Each motivational state competes with others to control the animal's behaviour – the one expressed behaviourally being the one in which the combination of causal factors is the highest. Thus, this model emphasizes how what an animal does depends on what other motivations are high, which in turn might be affected by what else there is to do in the environment.

A real example: a motivational model of dust bathing

In this chapter we have described several features of dust bathing in hens; which occurs in discrete bouts of about 20 min duration, most often in the middle of the day. Hogan and Van Boxel (1993) suggested a simple motivational model of dust bathing that captures these basic characteristics. They began by assuming a simple Lorenzian process whereby an internal factor for dust bathing builds up over time, but dissipates rapidly during dust bathing bouts (see Figs 4.7 and 4.8). The level of the factor

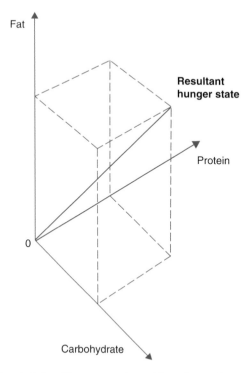

Fig. 4.7. The motivational state of hunger portrayed in a three-dimensional physiological space (modified from McFarland and Sibly, 1972, in McFarland, 1985).

G. Mason and M. Bateson

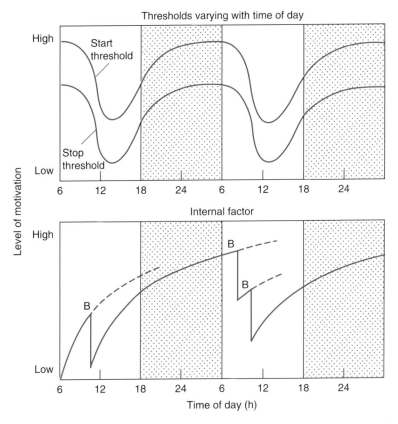

Fig. 4.8. A motivational model of dust bathing. The lower panel shows the internal factor for dust bathing, which is assumed to increase over time in a decelerating manner, and to decrease rapidly when a bout of dust bathing occurs (B). The upper panel shows the parallel upper and lower thresholds assumed to start and stop dust bathing, respectively. Dust bathing starts when the level of the internal factor exceeds the start threshold, and stops when it falls below the lower threshold. External stimuli such as light, heat and also qualities of the substrate are assumed to affect the thresholds (modified from Hogan and Van Boxel, 1993).

necessary for dust bathing to occur was assumed to depend on a threshold that is high in the morning but much lower in the middle of the day. If dust bathing continued until the internal factor returned to zero, then the model would predict dust bathing bouts to be longer in the morning than at midday (which is not observed). Therefore it was also necessary to assume a lower stopping threshold that varies in parallel with the starting threshold.

In ethology a good model of any kind will not only capture facts that we already know to be true about the control of a behaviour pattern, but will also make novel predictions that we can test. If the circadian variation in the height of the threshold is due to changes in light and temperature (a reasonable assumption to make), then we would predict that applying additional light and heat should shift the timing of dust bathing, but not the total amount performed. Hogan and Van Boxel (1993) were able to confirm these predictions in a study of Burmese red junglefowl. However, subsequently, other ethologists have argued that the above model does not capture some of the data on dust bathing, and have suggested alternative approaches (for more on this interesting topic see Olsson and Keeling, 2005).

4.4 Motivation and Applied Ethology

There are four key reasons for those interested in domestic animal behaviour to understand motivation, and we end our chapter with these.

Unfulfilled motivations and poor animal welfare

The first reason to understand motivation is related to animal welfare, and arises from the links between motivation and emotion. Emotions can be thought of as the states in an animal that are elicited by reward or punishment, whereas motivations can be thought of as states in which a reward is being sought or a punishment avoided (Rolls, 2005). Following these definitions, positive and negative emotions play an important role in the motivation of behaviours. In humans, for example, the affective (i.e. emotional) pleasure associated with drinking increases with motivation to drink, and the affective pleasure associated with touching something warm increases with motivation to be warm. This link between motivation and emotion led Dawkins (1990) to suggest that assessing motivational priorities was crucial for maximizing the welfare of animals kept in captivity: if captive animals prove highly motivated for resources they are denied in their typical housing conditions, then we should perhaps assume poor welfare due to negative emotions. All too often, captive animals have insufficient resources or opportunities for homeostasis, e.g. hunger (pregnant sows, broiler breeders, dairy calves), excessive cold/heat (mink underfed during the winter, cattle in shadeless feedlots, pigs transported in trucks) and other threats to homeostasis. When presented with opportunities to restore homeostasis, these animals show themselves highly motivated. For example, broiler breeds that are typically fed less than one-third of the ration they would choose to consume will work hard to obtain extra food from an operant schedule. Animals are not only motivated to restore homeostasis but, when asked – experimentally – if they will work for the chance to perform naturalistic activities, often answer 'yes', even when these behaviours offer no nutritional or other physiological benefit.

Thus, pregnant sows will work to build nests, as will hens about to lay. Mothers will often work for infant cues and, in some animals, same-age conspecifics are also valued. For example, isolated primates, rats and red fox vixens will all work hard for social company. Elements of natural foraging behaviours also seem highly motivated, even when they do not result in the actual acquisition of food. Thus rats, mice and many other animals will perform operant responses for the chance to run in a running wheel; the naturally semi-aquatic mink will work to swim; and pigs will work for the chance to root in bare earth or chopped straw. There are many other cases – examples that are poignant given that these animals are often denied these natural behaviours in captivity.

G. Mason and M. Bateson

Of course, to be sure that such data really demonstrate poor welfare in captivity, we need to check that these measured motivations are not simply elicited by cues from the resources to which these animals are exposed during these tests. However, in at least some of these examples (e.g. nest building for sows, social isolation for rats and primates and perhaps also swimming for at least some genetic subtypes of mink), there are good additional stress-related data (see Chapter 7) to show that preventing these natural behaviours in captivity is a real cause for concern. Natural behaviours, the prevention of which causes suffering, are termed 'ethological needs'.

Understanding stereotypic behaviour

The second reason to investigate domestic animals' motivations is that this can help us understand why abnormal behaviour patterns like stereotypic behaviours are so common (see Chapter 7). Stereotypic behaviours – abnormal repetitive behaviours with no apparent goal or function – are prevalent in captive animals: over 85 million perform them worldwide and, in some populations (e.g. zoo-housed giraffes and single-housed laboratory primates), they are nearly ubiquitous.

Ethological explanations for these behaviours have focused on frustration, since they often seem to arise from attempts to perform highly motivated normal behaviour patterns; in many instances, depriving animals of chances to perform behaviour by removing important resources induces or exacerbates stereotypies. Thus caged wild-caught birds typically show more route tracing than captive-bred birds, possibly because they are trying to escape; premature maternal separation typically induces similar stereotypic attempts to escape, or instead to suckle, in young mammals; and removing or delaying expected food rewards elevates stereotypic pacing, weaving and oral behaviours in captive pigs and carnivores.

The first biologists to notice and record these behaviours typically explained them as caused by internal states of deprivation (stemming from the loss of the mother, say, or from insufficient food), perhaps combined with strong eliciting stimuli (the sight of other conspecifics or the sounds and smells of other nearby animals being fed), leading to sustained high motivations that the subject cannot reduce: it cannot perform the activity that would result in negative feedback. Powerless to effect real change, the animal is able only repeatedly to attempt to perform a substitute for the relevant behaviour, or try and try again to escape the frustrating situation (see, e.g., Hughes and Duncan, 1988). The lack of motivational competition typical of barren captive conditions was also assumed to contribute to the prolongation of bouts of stereotypic behaviour.

Some of these motivational hypotheses have recently been tested experimentally. Important work in the 1980s on hunger and stereotypic oral activities in pregnant sows has already been extensively reviewed elsewhere. Here we present some newer and perhaps less familiar examples. For example, Wiedenmayer (1997) observing the stereotypic digging of caged gerbils, reasoned that, in the wild, the stimulus that would bring digging to a close would be a tunnel leading to a nesting chamber. He tested this idea in the laboratory and found that gerbils that had been raised with extra space, sand to dig in, or a simple plastic nesting chamber still showed the abnormal behaviour – but animals raised with a nesting chamber reached via a plastic tunnel did not. Nevison and colleagues (1999) similarly manipulated external stimuli to test the idea that bar-mouthing by laboratory mice derives from repeated escape attempts. They found that, if a door in the cage-lid was regularly opened and mice sometimes allowed to exit from it, their bar-chewing became directed to that site; on the other hand, if clear plexiglass was placed over a region of the cage-lid, to prevent odour cues from entering at that point, mice would move their bar-mouthing elsewhere.

More recent work still has focused on feather-pecking by hens, demonstrating that it derives from a very specific form of redirected foraging (rather than from redirected dust bathing pecks, as had been suggested in the 1980s). Harlander-Matauschek and colleagues (2007) discovered that individual hens who display a lot of feather-pecking are actually highly motivated to ingest feathers: offered bowls of food, shaving and feathers, they approach the feathers more rapidly, stay at the feather-bowls for longer and ingest more feathers than non-feather-pecking individuals. They will even perform an operant to gain feathers to eat. Meanwhile other researchers (Dixon *et al.*, 2008) examined the 'fixed action patterns' involved in feather-pecking: their careful quantitative comparison of the morphology of feather-pecks with foraging pecks, dust-bathing pecks, drinking pecks and exploratory pecks showed that foraging and feather-pecking involved near-identical fixed action-patterns.

These explanations, in terms of specific motivational frustrations, seem likely to explain the form and timing of many abnormal behaviours. However, they do not seem to be the whole story. There is growing evidence that these strange behaviours are exacerbated by impairments within the brain – induced by stress and/or by abnormal development – that predispose animals to pathological levels of inappropriate repetition. Some of these changes are quite independent of motivational affects, and simply add to them, or enhance them into the clockwork-like behaviours that are sadly so common (see Chapter 7). Some, however, may involve broader motivational abnormalities: generalized tendencies to respond, inappropriately excessively, to potential rewards perhaps related to the tendencies involved in human compulsions and addictions (see Chapter 7). This is an exciting area for future research.

Reward and motivation in learning and training

The third reason again relates to the links between motivation and emotion: that understanding motivation can increase the ease with which we can train animals to perform behaviour patterns that we want.

Motivation and learning are intimately linked: the opportunity to perform highly motivated behaviours is a powerful 'reinforcer' in learning – far more so than the opportunity to perform behaviours that are not highly motivated. Thus, in choosing, say, how to use food treats to reward desired behaviours during training, we need to understand that these will be effective only if the animal is motivated to obtain them. The use of unpalatable food rewards, or the animals being sated thanks to just having eaten a large meal, will therefore impede training, while in contrast the use of highly palatable treats, or prior moderate food deprivation, combined with the use of small treats that do not sate the animal, will all facilitate training by helping to ensure that the animal is sufficiently motivated to learn the behaviours we desire so that it can obtain the food rewards. This can have animal welfare implications too; rather than food-depriving rats or mice to get them to run tasks in the

G. Mason and M. Bateson

laboratory, researchers can choose to use highly palatable sweet breakfast cereals or condensed milk as rewards. Similarly, non-deprived hens and starlings will work for highly desirable mealworms.

Encouraging species-typical behaviour

The fourth and final reason is similar to the previous three in that, by appreciating how internal and external factors combine, in an unlearnt way, to motivate behaviour, it can help us 'persuade' the animals we keep to behave as we humans want them to.

For example, if we appreciate how dam endocrine state and lamb odour combine to motivate maternal care in sheep, we can use this information to devise the best ways to manipulate ewes so that they will accept orphan lambs (by timing our attempts appropriately relative to parturition; and by modifying olfactory cues from the orphaned lamb). We can use our knowledge that highly motivated animals will be less choosy about how to satisfy their motivations to persuade sexually aroused stallions to ejaculate into an artificial mare for semen collection (by first using a 'teaser' mare, in oestrus, to arouse the stallion); or similarly to persuade females to mate with sub-standard males. Conversely, we can appreciate that, if internal factors motivating, say, eating are low, we will have to use particularly palatable foodstuffs – i.e. treats that present particularly powerful eliciting external stimuli – if we want to persuade ill or frightened animals in our care to eat.

4.5 Conclusions and Links to Other Chapters

In this chapter we have provided a brief introduction to the data and theories relating to the concept of motivation, and have discussed these in the context of some of the current challenges in applied ethology. We would like to end by returning to Tinbergen's four questions and emphasizing the strengths of the ethological approach. Although the study of motivation focuses on questions of proximate causation, in trying to understand the motivation for a particular behaviour pattern it will often be important to consider Tinbergen's other three questions (see Chapter 1). The causal mechanisms responsible for a behaviour pattern are often affected by the developmental experience of an animal, and thus a good understanding of the ontogeny of a behaviour pattern could be crucial in understanding why a captive animal is behaving abnormally. The causal mechanisms responsible for a behaviour pattern have evolved in order to motivate an animal to perform behaviour patterns that will optimize its survival and reproduction. Therefore, understanding the ultimate goal and survival value of a behaviour pattern, and its phylogenetic roots, is also likely to be important in understanding how it is controlled. This is particularly important when solving problems in applied ethology because, as we have seen, the unfulfilled ethological needs and abnormal behaviour patterns of captive animals may often have their origins in adaptive behaviour.

Acknowledgements

Thanks to Laura Dixon and Naomi Latham for some useful examples, and to David Fraser for Fig. 4.1.

References

Barnard, C. (2004) *Animal Behaviour: Mechanism, Development, Function and Evolution.* Pearson, Harlow, UK.

Berridge, K.C. (2004) Motivation concepts in behavioral neuroscience. *Physiology and Behaviour* 81, 179–209.

Boulton, M.I., Wickens, A., Brown, D., Goode, J.A. and Gilbert, C.L. (1997) Prostaglandin F2 alpha-induced nest-building in pseudopregnant pigs. 1. Effects of environment on behaviour and cortisol secretion. *Physiology and Behaviour* 62, 1071–1078.

Cooper, J.J. and Mason, G.J. (2000) Increasing costs of access to resources cause rescheduling of behaviour in American mink *Mustela vison*: implications for the assessment of behavioural priorities. *Applied Animal Behavior Science* 66, 135–151.

Dawkins, M.S. (1990) From an animal's point of view: motivation, fitness and animal welfare. *Behavioural and Brain Sciences* 13, 1–61.

Dixon, L, Duncan, I.J.H. and Mason, G.J. (2008) What's in a peck? Using fixed action patterns to identify the motivation behind feather-pecking. *Animal Behaviour* 76, 1035–1042.

Harlander-Matauschek, A., Benda, I., Lavetti, C., Djukic, M. and Bessei, W. (2007) The relative preferences for wood shavings or feathers in high and low feather pecking birds. *Applied Animal Behaviour Science* 107, 78–87.

Hogan J.A. and Van Boxel, F. (1993) Causal factors controlling dust bathing in Burmese red junglefowl: some results and a model. *Animal Behaviour* 46, 627–635.

Hughes, B.O. and Duncan, I.J.H. (1988) The notion of ethological 'need', models of motivation and animal welfare. *Animal Behaviour* 36, 1696–1707.

Mason, G., Cooper, J. and Garner, J. (1997) Models of motivational decision-making and how they affect the experimental assessment of motivational priorities. In: Forbes, J.M., Lawrence, T.L.J., Rodway, R.G. and Varley, M.A. (eds) *Animal Choices*. Occasional Publication No. 20, British Society of Animal Science, Penicuik, UK, pp. 9–17.

McFarland, D. (1985) *Animal Behaviour.* Longman Scientific and Technical, Harlow, UK.

Nevison, C.M., Hurst, J.L. and Barnard, C.J. (1999) Why do male ICR(CD-1) mice perform bar-related (stereotypic) behaviour? *Behavioural Processes* 47(2), 95–111.

Olsson, I.A.S. and Keeling, L.J. (2005) Why in earth? Dust bathing behaviour in jungle and domestic fowl reviewed from a Tinbergian and animal welfare perspective. *Applied Animal Behaviour Science* 93, 259–282.

Ramsay, D.S., Seeley, R.J., Bolles, R.C. and Woods, S.C. (1996) Ingestive homeostasis: the primacy of learning. In: Capaldi, E.D. (ed.) *Why We Eat What We Eat.* American Psychological Association, Washington, DC.

Rolls, E.T. (2005) *Emotion Explained.* Oxford University Press, Oxford, UK.

Warburton, H. and Mason, G. (2003) Is out of sight, out of mind? The effects of resource cues on motivation in mink, *Mustela vison. Animal Behaviour* 65, 755–762.

Wiedenmayer, C. (1997) Causation of the ontogenetic development of stereotypic digging in gerbils. *Animal Behaviour* 53, 461–470.

Wiepkema, P.R. (1971) Positive feedback at work during feeding. *Behaviour* 39, 2–4.

5 Learning and Cognition

M. Mendl and C.J. Nicol

5.1 Introduction

A young sow approaches a site on the boundary between forest and open field where she had found some tasty roots the previous day. She is poised to start rooting the ground when she notices one of her group-mates, a large dominant female, approaching. Instead of digging for the roots, she turns away and continues along the forest boundary. Close by, a field ethologist, having studied the behaviour of this group of sows for several weeks, scribbles in his notepad. The young sow appears to have remembered where the good food source was, something that he had observed previously, but why did she not feed this time? Is it possible that she didn't want to reveal the location of the source to the approaching dominant sow? If so, wouldn't this suggest that she had some understanding of what would happen if she gave away her secrets?

This is not just a fanciful example; pigs show intriguing behaviour like this in controlled experimental studies (e.g. Held *et al.*, 2002). But it raises a number of important questions about the processes underlying the behaviour that we observe. How does the sow remember where food was? Does she form an association between a particular visual stimulus, a clump of trees, say, and food? Or does she have a complete mental representation of the area? Why does she not feed? Assuming that she recognizes her dominant group-mate, does she associate the dominant with previous painful skirmishes over resources and simply avoid her? Or does she actually understand what will happen if she started to feed, and perhaps even attribute food-pilfering intentions to the dominant sow?

These questions reveal that apparently 'clever' behaviour can be explained by a variety of putative underlying mental or 'cognitive' processes. These range from associative learning through to the capacity to form 'cognitive maps' or to have a 'theory of mind', abilities that involve quite complex mental representation of the outside world. Distinguishing between these explanations for observed behaviour is often not easy and is a theme that we will return to. However, our principal aim in this chapter is to consider the evidence for a number of different types of cognitive ability in domestic animals.

Before we go any further, what exactly do we mean by 'cognition'? Historically, the behaviourist school of psychology that dominated the study of animal learning for much of the 20th century, through the work of J.B. Watson, B.F. Skinner and others, sought to explain all behaviour without referring to unobserved mental processes. Indeed, it denied that it was possible to study these at all. However, the cognitive revolution of the latter part of the 20th century has re-legitimized the study of hidden mental processes through careful inference from observed behaviour. Consequently, for some authors, cognition refers to these mental processes, and the study of cognition seeks to understand how information is represented and manipulated in the mind, in stark contrast to the behaviourist approach. Others, however, use the term in a broader sense to refer to all processes by which 'animals acquire, process, store and act on information from the environment' (Shettleworth, 1998).

In this chapter, we discuss a number of examples of animal cognition (*sensu* Shettleworth) but, where relevant, also consider the extent to which they can be explained by behaviourist and cognitivist theories. We should also emphasize that, when we refer to 'mental representations', we do not imply that these are consciously experienced by non-human animals in the way that we humans experience a thought or a feeling. This may be the case but at present we do not know. It is important to remember that the question of whether animals are consciously aware of their mental processes is distinct from the question of how these processes work. The study of cognition is concerned with the latter question only.

Learning and cognitive abilities should enhance the inclusive fitness of an individual, increasing its chances of contributing genetically to the next generation. Different aspects of learning will be of adaptive benefit to different species, making it very difficult to say whether one species is 'more intelligent' than another, although the capacity to learn will certainly be influenced by environmental and social complexity. Animals born into constant environments can survive well using innate responses but, when environments are varied and unpredictable, a wide range of learning abilities will be essential. The ancestors of our most common domestic mammals and birds will have faced many uncertainties, including food of variable quality or unpredictable distribution, predators with different appearances, location and habits, and complex social environments where the identities and roles of group members vary over time. This varied heritage means that domestic animals should easily be able to establish relationships between events and thus form associations that guide behavioural change.

We start by considering the associative learning processes underlying these abilities. We then consider memory processes, including the special example of object permanence, before discussing discrimination, generalization and category formation, social cognitive abilities and the relationship between cognition and emotion. Despite a long domestication history, we will see that domestic animals show surprising mental flexibility and complexity in the way that they interact with the environment and their social companions. Unfortunately, space constraints mean that we cannot address other topics. However, those that we do cover have relevance for domestic species, and we conclude the chapter by briefly discussing how an understanding of animal cognition influences our attitudes towards and understanding of animal welfare.

5.2 Associative Learning

Two main types of associative learning have been described and extensively studied. In classical, or Pavlovian, conditioning, an environmental event or stimulus is followed predictably by some other occurrence. Pavlov studied the effects of sounding a bell just prior to food arrival on salivation responses in dogs. The dogs naturally salivated to the smell or appearance of food, so salivation was described as an unconditioned response (UR) and food as an unconditioned stimulus (US). However, after repeated pairings of the bell with the food, the dogs salivated in response to the sound, i.e. the conditioned stimulus (CS). When an association is classically conditioned animals do

M. Mendl and C.J. Nicol

not acquire new responses or behaviours, so the conditioned response (CR) takes the same form as the UR. Such associations are acquired with ease. Companion dogs, for example, can become expert at detecting the subtlest signs that might predict their daily walk, without any formal training. The sight of a human picking up a lead (the CS) can set off a volley of anticipatory responses such as barking and leaping. Classical conditioning therefore allows animals to *predict* events, but it gives them little control or influence over the actual timing, quality or duration of their meals or walks.

Instrumental, or operant, conditioning is a second type of associative learning, whereby an animal directs a behaviour to a new part of its environment or learns to perform a new behaviour, to obtain a reward. Imagine, for example, a captive zoo-housed coati (*Nasua nasua*), presented for the first time with a cylindrical environmental enrichment device containing small food pellets that fall through holes only when the device is manipulated and moved in a certain way. The coati might initially approach the cylinder and perform a range of natural exploratory movements (URs). But, if one particular movement results in the acquisition of a food pellet (US), the coati will repeat the previous few movements it made, until it learns precisely which action (now a conditioned response, CR) reliably produces food. In this case, the animal gains not only the ability to *predict* what might happen next, but also the ability to *control* the timing and delivery of its own reward.

An influential model providing insight into the process of association formation was outlined by Rescorla and Wagner in 1972 (reviewed by Hall, 1994). This model describes many features of associative learning, although it provides no insight as to the neurological mechanisms involved. Repeated pairings of events result in associations with strengths that reflect the exact statistical probability that a CS (e.g. a bell sounding) or a CR (e.g. an action directed at a cylinder) will result in US arrival. The *contiguity* between two events is therefore more important in establishing an association than the absolute temporal relation between them. Usually, if a stimulus or response is followed rapidly by US arrival, a good association will be formed, because there is no time for intervening events to interfere with the predictive relationship. However, under some conditions, an association can be formed even if the CS precedes the US by many hours, provided the relationship between the two is statistically strong.

The best examples of delayed association formation come from ecologically appropriate situations. In food aversion learning, for example, an animal can ingest a novel (slightly toxic) substance and not feel sick for many hours. Once nausea sets in, however, it will be able to associate the sickness with the novel substance it ingested, rather than with any of the sights or sounds that it encountered in the intervening period. The novel food is biologically the most plausible predictor of sickness. The Rescorla–Wagner model makes a number of other predictions that have been valuable in explaining empirical data. For example, association 'strength' depends on factors such as the size or importance of the US, and the presence of existing associations (e.g. the bell that predicts food) can 'interfere' with the formation of new associations (e.g. that a light also predicts food) via processes called overshadowing or blocking (Dickinson, 1980).

Associative learning takes place only when an unconditioned stimulus is biologically meaningful: something the animal either wants, or wants to avoid. The unconditioned stimulus is often called a reinforcer. Some reinforcers result in *increased* performance of the behaviour under consideration. Appetitive reinforcers can be positive, where the animal responds to gain more of the reinforcer (e.g. food), or negative, where the animal responds to avoid the reinforcer. Physical pressure, used in horse training, is a good example of a negative reinforcer. Horses will move forwards or sideways to obtain the reward of removal of the rider's leg pressure. In contrast, reinforcement that produces a *decrease* in the behaviour under consideration is known as punishment. For example, if a dog stops barking when it is shouted at, then shouting is a punishment for that animal. But there is nothing beyond an animal's own reaction that can be used to define a situation as good or bad in advance. A second dog may like its owner shouting and bark more whenever it is shouted at. For this second dog, barking is a positive reinforcer and not a punishment, whatever the intentions of the owner!

Other rewards such as social contact or neutral thermal conditions may be important in guiding learned behaviour, but they are difficult to use in experimental or training situations. Because of this, conditioned or secondary reinforcers, initially neutral but paired with a biologically meaningful stimulus, are often used when training animals. Interest in this area is exemplified by articles and books on 'clicker training' in the popular press. However, experimental studies of the efficacy of secondary reinforcers have not always shown that they promote the faster learning or greater retention sometimes claimed. Williams *et al.* (2004), for example, found that conditioned reinforcement did not improve learning acquisition, or resistance to extinction, in horses.

Associative learning: what is learned?

Studies of associative learning were the bedrock of behaviourist research and, in many cases, it was possible to explain changes in behaviour in terms of stimulus–response connections that required no reference to unobserved mental processes. However, this was not always so. For example, rats exposed to the coincidence of two biologically neutral stimuli – 'tone predicts light' – showed no change in behaviour, indicating to a behaviourist that nothing had been learnt. However, if the light then became predictive of electric shock, the rats subsequently showed avoidance behaviour to the tone as well. This could be explained by postulating that the rats formed mental representations of the relationships between tone, light and shock and were thus able to combine the information 'tone predicts light' and 'light predicts shock' (see Dickinson, 1980). A cognitivist perspective, therefore, accounted for the observed behaviour.

The cognitivist view further postulates that information acquired during associative learning can be stored in two different ways. *Declarative* representations, as in the above example, involve knowledge about things or relationships (e.g. the light predicts shock) that allow different representations to be combined in different ways, and confer flexibility in how the animal uses the information to guide a behavioural response (Dickinson, 1980). *Procedural* representations, on the other hand, involve knowledge about what to do (e.g. when the light is on, crouch), as in a simple stimulus–response connection, and can be used only in an inflexible way. There is some evidence that chickens may be able to integrate information about the location of a particular food type in a foraging arena, and the value of that food type, and use this information to guide their choices, suggesting that they may form declarative representations (Forkman, 2001). If so, chickens and, very likely, other domestic species may be capable of more sophisticated forms of associative learning than the acquisition of simple stimulus–response relationships.

5.3 Memory

From a cognitivist perspective, associative learning involves changes to representations of information and the storage and retention of these – memory. Studies of learning and memory thus overlap, and to a certain extent the two phenomena are inseparable. However, memory research can be characterized by the types of question that it addresses, which include: (i) how is information acquired, stored and retrieved; (ii) for how long can information be retained; (iii) how large is the memory store; and (iv) what are the neural substrates of memory?

Memory research has also identified a number of different types of 'memory system'. Working memory usually refers to the capacity to hold information in some form of cognitive workspace for a very short time (e.g. seconds). In humans, it has been suggested that a maximum of around seven pieces of information (e.g. seven numbers) can be held in working memory. In animal studies, the terms working memory and short-term memory are often used interchangeably to describe the storage of information over a few minutes to hours, as in when an animal is foraging for food and uses working memory to avoid revisiting locations that it has just searched. Long-term or reference memory holds information for much longer periods (days, months, years) and seems to have a virtually unlimited capacity. For example, there is evidence that some food-storing birds can remember the location of thousands of food-cache sites. Long-term memory also seems to require significant molecular and cellular events, including protein synthesis, in areas of the brain that are implicated in memory processes, such as the hippocampus, amygdala and medial temporal lobes. Although far from being fully understood, these can be conceptualized as the laying down of a 'memory trace'.

Working, short- and long-term memory abilities have been demonstrated in a variety of domestic species, often using ecologically relevant spatial tasks such as remembering the location of food. Pigs show a well-developed spatial memory; they can be readily trained to search an arena each day for food in a novel location, retain that memory over the next 1–2 h and revisit the same (re-baited) location when allowed back into the arena. They demonstrate good working memory in that they usually avoid returning to previously visited locations during the few minutes of each search of the arena, and this is not due to following a simple 'turning rule' (e.g. always moving to the left of the site that has just been visited). They easily remember that day's correct location over a 2 h retention interval, and improving performance over consecutive days indicates that they learn that the first search of each day involves finding the new location of food, while the second search involves remembering where food was on the first search, thus indicating good reference or long-term memory of what the task is all about (Mendl *et al.*, 1997).

Cattle and sheep also perform well in spatial memory tasks. For example, cattle can develop good long-term memory of where high- or low-quality food is in a radial-arm maze, showing a preference for high-quality sites and avoiding low-quality ones. In the case of high-quality food, this memory persisted for at least 30 days in one study, while memory-aided avoidance of low-quality food locations appeared to decline during the same period (Bailey and Sims, 1998). Sheep can also develop good long-term memory of the location of high-yielding areas in large arenas (e.g. 160 m × 160 m), and this ability allows them efficiently to exploit different quality pastures (Dumont and Petit, 1998).

These examples of spatial memory probably involve associative learning processes that link visual, olfactory or other spatial cues with food reward. Distant visual cues are often used by animals to orient themselves and solve these tasks, but there is also evidence that cattle and sheep can use visual cues close to the food source itself to guide their searching. Whether these species can develop 'cognitive maps', mental representations of the spatial environment that allow them to make novel short cuts, remains to be discovered.

Another ecologically relevant ability, memory for other individuals – or at least for cues from those individuals such as facial appearance – has also been studied in domestic species. One study demonstrated that sheep could discriminate between pairs of conspecific faces as evidenced by learning to associate one member of each pair with a food reward. They also appeared to be able to generalize from frontal views to profiles, and they were able to perform accurately on these discriminations after 2 years, although there was a decline in performance, and it is possible that they were mainly remembering the task and could have performed equally well on new face pairs (this was not reported). Neurophysiological evidence indicated that neurons in the temporal and medial prefrontal cortex of the brain, which fire when sheep view a familiar conspecific, continued to fire in response to specific facial images, even after 8–12 months in which that individual had not been encountered (Kendrick et al., 2007).

While the capacity of domestic species to store information in memory is not in doubt, the variables that affect the selection, storage and retrieval of memory have received less attention, but are critically important in determining what is remembered and how well (see Fig. 5.1). Clearly, we don't remember everything that happens to us. In fact, if we did, it might drive us crazy. There is increasing evidence that, just as biologically meaningful unconditioned stimuli trigger associative learning, the events that are prioritized for storage in memory are also 'important' ones, those which are most likely to impinge on survival and reproductive success and, probably through 'evolutionary design', also have emotional impact. In support of this idea, moderate elevations of 'stress hormones', such as cortisol or adrenalin, as would probably occur in response to significant events, can aid consolidation of information into memory, probably through direct or indirect action on brain structures such as the amygdala and hippocampus. Furthermore, a study of pigs indicates that information that is more costly to forget, in this case because forgetting it increases the time taken to obtain food, is indeed encoded more effectively in memory (see Mendl et al., 2001).

Information stored in memory may 'decay' with time, in that it is less readily retrieved as time passes. One process that may contribute to this is termed retroactive interference and occurs when new information presented to the animal apparently interferes with the stored representation of similar information. For example, pigs that have been trained to remember the location of food in an arena appear to have this memory disrupted if they are exposed to the arena again (without being allowed to search it) prior to being tested for memory retrieval (Mendl et al., 2001). One explanation for this is that the memory trace of the original information is 'reactivated' from a 'dormant state' when the animal encounters a relevant cue. Once reactivated, the

M. Mendl and C.J. Nicol

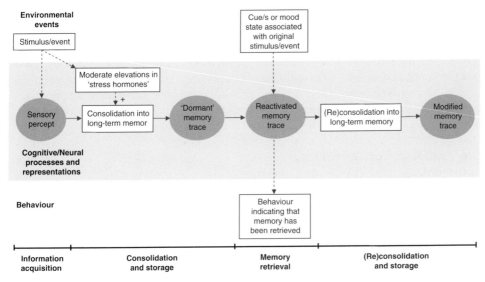

Fig. 5.1. Schematic representation of some of the events that are thought to occur when information is acquired, stored and retrieved in memory. A stimulus or event is perceived and a representation of this is consolidated into long-term memory as a neural 'memory trace'. This involves molecular processes, including protein synthesis in cells, and can be facilitated by moderate elevations in 'stress hormones', ensuring that important or emotionally salient events are given priority in memory. When a cue related to the original stimulus and/or a mood state associated with the original stimulus is encountered, the 'dormant' memory trace appears to be reactivated, as evidenced by behaviour indicating that the memory has been retrieved. New information present at this time may then be incorporated into a newly consolidated memory trace, which may thus be slightly, or even greatly, altered from the original memory.

memory trace is then 'labile' and vulnerable to alteration by new information, before it is stored again in a dormant form. It is possible to investigate this idea because the (re)storage of longer-term memories requires protein synthesis in specific brain areas, and this can be temporarily stopped using protein synthesis-blockers.

Using such techniques it has recently been shown that a ewe's memory trace of her lamb does indeed appear to be reactivated by cues from the lamb, and can then be disrupted such that her subsequent ability to recognize the lamb deteriorates (Perrin *et al.*, 2007). These and other findings emphasize the complexity of memory processes (see Fig. 5.1) and help to explain how memories can be interfered with and decay, and also how human memories can change and become embellished with time.

Retrieval of memories can be triggered by presentation of appropriate cues, such as the context in which an event took place or the smell of another individual, and there is some evidence that information with a particular affective value (e.g. anxiety-inducing) is more readily retrieved when the subject is in a congruent state (see Fig. 5.1). There has been limited research on how retrieval is affected by these variables in domestic species. However, one study of laboratory rats found that short-term memory for another individual appeared to be unaffected by whether the individual, or a representative odour cue, were re-encountered in the same context as originally or in a different, but familiar, context (Mendl *et al.*, 2001). Whether animals can retrieve

---

memories in the absence of an overt cue, so-called 'recall' in contrast to cue-driven 'recognition', is of course very difficult to test. Furthermore, exactly what is retrieved remains mysterious. Whether animals have vivid conscious recall of events that happened to them in the past, termed 'episodic memory' in humans, remains unknown and perhaps unknowable, although there is increasing evidence that at least some species, including laboratory rats, may be able to remember not just what happened where, but also when it happened.

5.4 Object Permanence

Object permanence is the awareness that objects are separate entities that continue to exist when out of sight of the observer, and can be thought of as a specific type of memory. The development of this awareness in humans was studied by Piaget, who noted that babies of less than about 8 months of age sometimes behaved as if an object no longer existed when it was hidden behind a screen (reviewed by Pepperberg, 2002). These infants did not try to search for the object and behaved as if 'out of sight' was 'out of mind'. In contrast, older babies would search for a toy or other object obscured by a screen or a blanket. Piaget suggested that object permanence became increasingly sophisticated as children aged, culminating in the ability to infer where objects were, even after a series of invisible displacements. Improved techniques for studying human babies have shown that they may develop these abilities much earlier than originally thought.

In the light of these improved techniques, it is interesting to consider whether domestic animals have object permanence and, if so, at what level of complexity. The developmental sequences described for humans are unlikely to be relevant to precocial species such as horses or chickens, but object permanence can still be examined within a comparative framework. Parrots can accomplish sophisticated invisible displacement tasks (Pepperberg, 2002), whilst dogs (Fiset et al., 2007) and chickens (Freire et al., 2004) are adept at finding objects that have been hidden inside or behind visible containers or screens, even if they have to take detours or move out of sight of the container as part of the retrieval process. The ability to retrieve a hidden object depends not only on some form of awareness that it still exists, but on developmental experience with occluded objects (Freire et al., 2004) and on a variety of spatial memory and navigation processes (see Fig. 5.2).

5.5 Discrimination, Generalization, Categorization and Concept Formation

Discrimination allows distinctions to be drawn between objects or stimuli that differ in particular features. There appear to be natural biases in the ease with which different species attend to and use different stimulus features for discrimination, which almost certainly depend on adaptive ecological history. In tests where both the appearance and relative position of buckets are potential cues indicating food availability, horses preferentially utilize the positional cues (Hothersall et al., 2009). In chicks the situation is more complicated, in that the right brain hemisphere preferentially attends to position-specific cues, whereas the left hemisphere attends to object appearance (Regolin et al., 2004).

M. Mendl and C.J. Nicol

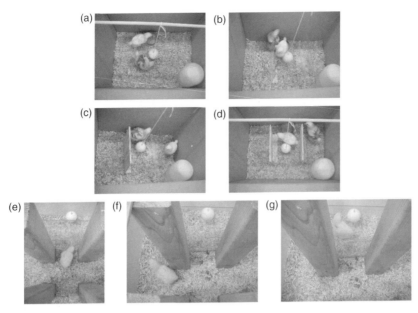

Fig. 5.2. Experience of occlusion of objects at around day 11 of life is important in the development of object permanence abilities in chicks. Freire *et al.* (2004) kept chicks from 8 days of age under conditions that permitted different opportunities to move out of sight of a yellow tennis ball on which they had previously been imprinted. Chicks were housed in pens with either: (a) no screens, (b) two transparent screens, (c) one opaque and one transparent screen or (d) two opaque screens. Chicks with opaque screens spent most time out of sight of the ball, and subsequently made fewer errors in locating the ball in a detour test (e, f, g).

Discrimination also enables animals to fine-tune their behaviour in response to environmental cues. A dog may learn that barking results in attention from the owner when the TV is off, but not when the TV is on. The status of the TV therefore controls the behaviour of the dog, much as a laboratory rat might discriminate by performing an operant response in the presence of one stimulus but not another. Generalization is the converse tendency to attend to shared features within a range of stimuli. Generalization enables animals to respond to new stimuli adaptively, provided sufficient shared information exists with previously encountered stimuli. The adaptive balance between discrimination and generalization depends on precise circumstances. A riding school pony, for example, has to respond appropriately to a variety of riders using leg, seat and hand cues in different ways. The adaptive response of the pony is to generalize and respond in the same way to a broad range of different cues. For an advanced dressage horse, such generalization would be disastrous and dressage riders aim for precision and consistency in their aids in order to foster discrimination and avoid generalization.

If the shared features of a class of stimuli are based on physical similarities such as colour, size or shape, then the processes of discrimination and generalization can result in category formation. Horses can categorize trained and novel images according to whether they have open centres or are solidly black, when procedures are used that control for confounding variables such as the overall areas of black and white in the stimuli (Hanggi, 1999).

Categorizing stimuli according to abstract principles such as properties that result from *relationships* between stimuli, e.g. 'bigger than' or 'paler than', is cognitively more demanding. Well-designed experiments have demonstrated that primates, parrots and pigeons can form such relational categories, but there is limited information about the abilities of domestic animals on these tasks. The only relational category that has been studied to any real extent in domestic animals is the 'same/different' category. This can be examined using matching-to-sample or non-matching-to-sample tests, where an animal capable of this degree of abstraction should be able to select a stimulus that is the 'same as' or 'different from' one previously viewed. Same/different categorizations have been difficult to demonstrate in some species, but studies using olfactory stimuli in rats and auditory stimuli in dogs have recently shown more success, and such methods could fruitfully be adapted to examine abstract categorization abilities in other domestic animals.

An entirely abstract category could be called a 'concept' or an 'idea of a class of objects', where stimuli are included on the basis that they stand for the same idea and not that they resemble each other physically in any way. Concept formation, especially in animals that have no language, is of great theoretical interest but few studies have been conducted with domestic animals, and developing viable methodologies remains a challenge (Lea *et al.*, 2006).

5.6 Social Cognition

For many species, the social environment is highly complex and changeable, and some of the most interesting examples of cognitive abilities arise in this context. Notably, social learning facilitates the acquisition of new behaviours in many domestic animals. Social learning occurs when a naive animal (the observer) acquires information from a knowledgeable conspecific (the demonstrator), resulting in faster or more effective learning by the observer, often with greatly reduced costs. Sometimes, the demonstrator simply draws the observer's attention to a previously unnoticed stimulus, and subsequent acquisition of a new response by the observer behaviour then takes place by normal instrumental conditioning. But social learning can also occur via processes that are cognitively more demanding, including imitation, where an animal copies the physical movements of a demonstrator; and emulation, where it reproduces the results of a demonstrator's actions (Whiten *et al.*, 2004).

There is good evidence that most domestic animals employ social learning in some situations, but the processes involved have perhaps been most studied in chickens (reviewed by Nicol, 2004), where, in early life, chicks are sensitive to social guidance about which foods they should and should not ingest. Hens attract their young to food with a complex display of staccato food calls and pecks to the ground. The hens' display is more intense in the presence of high-quality food items, if the chicks move away or if the chicks make apparent 'errors' in the objects they peck at. Older chickens will learn to eat, but not to avoid, novel-coloured food after watching a knowledgeable demonstrator. The relationship between the observer and demonstrator birds is very important, as hens are more likely to acquire new behaviours after watching socially dominant demonstrators than social subordinates.

As well as transmitting information about food availability or quality, animals communicate with each other about predator presence and identity, and about their own internal reproductive or emotional state (Manteuffel *et al.*, 2004). Traditional

M. Mendl and C.J. Nicol

texts hint that animals such as chickens might be rather inflexible in their communication abilities but, as seen by the responses of hens to the feeding behaviour of their chicks, communication can be both subtle and flexible. Another example comes from the alarm-calling behaviour of cockerels, who give different types of calls in response to perceived aerial or ground predators, and modulate their calling depending on the type of 'audience' that may be listening. Cockerels are most keen to impress novel female chickens and call more in their presence than in the presence of familiar females or novel birds of a different species. In addition, both male and female (broody) bantam chickens adjust their alarm calling according to the size of the aerial predator relative to the size of their own growing chicks. Alarm calls were given only in response to small hawks when chicks were very young and vulnerable (Palleroni et al., 2005).

Complex and flexible communication does not, however, mean that domestic animals possess language abilities. Language requires that new meanings can be generated by an almost infinite rearrangement of discrete symbols (such as words), or from rearrangements of symbol order (as in grammar). There is considerable debate about the extent to which symbolic use in chimpanzees and parrots meets these (admittedly anthropocentric) criteria for 'language', and there is no hint of current evidence that any domestic species possess such specialized abilities.

Perhaps the most advanced form of social cognition is the ability to 'put oneself in another's shoes' – for example, to take their visual perspective, to empathize with them, to understand that they have a mental state like oneself and even to use this knowledge to deceive them. Not surprisingly, it is very difficult to study these abilities involving high-level mental representations, and also to be sure that any behavioural evidence for them cannot be explained by 'simpler' associative learning processes. Primatologists research, and argue about, this area, but there have been only a few studies in domestic animals. As in the example at the start of this chapter, pigs can show behaviour that allows them to minimize exploitation of their knowledge of where food is in a foraging arena by another individual (see Fig. 5.3). However, although this may indicate a sophisticated understanding of the intentions of the other animal, it is more likely that this 'clever' behaviour is the result of the knowledgeable animal avoiding a competitive encounter with its larger companion over food, having learned by association that it is likely to lose out (Held et al., 2002).

One technique designed by primatologists to reveal more precisely what one animal understands about another's knowledge has been adapted for pigs and dogs. Here, the subject animal chooses to use one of two individuals as a guide to where the food is. The subject is unable to see where food is being placed in an arena, but can see that one of the individuals (the 'knower') has visual access to this baiting event while the other (the 'guesser') does not. The question is whether the subject then uses information provided by the 'knower' to locate food, which would indicate some understanding of the relative knowledge of the two individuals. One out of ten pigs made this 'correct' choice (Held et al., 2001), while no dogs did when tested with conspecifics, though they did when tested with human partners (Cooper et al., 2003). Despite well-designed experiments, it is still difficult to rule out the possibility that animals showing the correct response were doing so on the basis of previously learned associations rather than by taking the visual perspective of the other individual. For example, they may have learned that another's gaze direction or visual access to an event is a good predictor of subsequent behaviour. Dogs, at least, are highly sensitive to gestures and even the glancing behaviour of humans.

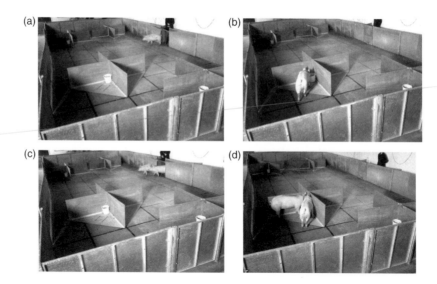

Fig. 5.3. Pigs undergoing a competitive foraging task. The animals are trained and tested daily in a foraging arena. (a) On the first trial of the day a pig enters the arena to search for food placed randomly in one of eight buckets; (b) it finds the food and eats it. It has previously been trained that, on the next trial a few minutes later, food will be in the same location as on the first trial; (c) the 'informed' pig is now introduced to the arena with a heavier companion, which has been trained that there is food in the arena, but does not know where it is; (d) if the companion follows the informed pig to the correct bucket, it is able to displace it and steal the food. Over many days it is possible to observe how the behaviour of the two pigs changes as first the companion learns to follow the informed pig to the food and then the informed pig develops behaviours which minimize the chances of the companion pig stealing its food (see Held *et al.*, 2002).

5.7 Cognition and Emotion

One recent area of interest has been in the use of cognitive measures as potential indicators of emotional states in animals. Emotions are states like happiness, sadness, fear and anxiety. Understanding these states in animals is critical to a better understanding of animal welfare. We consciously experience emotions, but we cannot be sure that animals do. Nevertheless, emotions are also accompanied by behavioural and physiological changes (e.g. an urge to flee, raised heart rate), and we can use these as proxy indicators of animal emotional states. The problem is that these measures may not always reliably reflect whether an emotion is positive or negative ('valence'), and are often species specific and therefore difficult to generalize a priori to other species. There are also few measures of positive emotions (Paul *et al.*, 2005).

Human studies have shown that cognitive processes may both influence and be influenced by emotional states, suggesting that they may be useful and novel indicators of animal emotion. For example, when an event happens there is evidence that people perform an extremely rapid series of stimulus checks (e.g. how familiar is it? How predictable is it? How sudden is it?), and the outcome of these 'cognitive appraisals' influences the emotion that is experienced. Particular emotions appear to be associated with specific stimulus characteristics, and it is possible that animals exposed to such

M. Mendl and C.J. Nicol

stimuli may also experience similar emotions. If so, the behavioural and physiological responses they show to these stimuli will provide good indicators of the corresponding emotion. Current work with sheep is investigating this possibility and is showing, for example, that stimuli differing in suddenness and novelty do indeed produce different behavioural and physiological response profiles (Boissy *et al.*, 2007).

Just as cognitive appraisals may influence emotional state, so emotional state affects cognitive processes including attention, memory and judgement. For example, people in a negative affective state are more likely to interpret an ambiguous event negatively, and to anticipate negative rather than positive things happening (Paul *et al.*, 2005). These emotion-related 'cognitive biases' have been explored in animals using discrimination learning protocols in which subjects are trained to make response A to stimulus X in order to acquire a good thing (e.g. food), and response B to stimulus Y (in the same sensory modality as X) in order to acquire a less good thing (e.g. less food) or avoid a bad thing (e.g. noise). Once trained, the subjects are presented with ambiguous stimuli (intermediate between X and Y) to see whether they perform response A, indicating that they categorize the stimulus as predicting a good thing ('optimistic'), or response B, indicating a 'pessimistic' judgement of the stimulus (see Fig. 5.4). The prediction is that subjects in a putative negative affective state are more likely to show 'pessimistic' responses. Results indicate that this is the case in species including rats, starlings and dogs (e.g. Harding *et al.*, 2004), raising the possibility of using cognitive bias as a novel indicator of animal emotion.

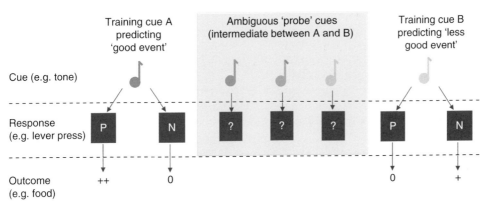

Fig. 5.4. One example of a procedure for a cognitive bias task. In this 'judgement bias' task, the animal is trained that when training cue A (e.g. a tone of a specific frequency) is presented it must perform response P to acquire a good reward (e.g. several pellets of food), while when training cue B (e.g. a tone of a different frequency) is presented it must perform response N to acquire a less good reward (e.g. one pellet of food). Once trained to this task, it can be assumed that performance of response P indicates anticipation of a good event, whilst performance of response N indicates anticipation of a less good event. The animal is then presented with ambiguous cues (tones of intermediate frequency) and its responses to these are recorded. The prediction is that animals in a negative mood state are more likely to categorize the ambiguous cues as predicting a relatively negative event and hence are more likely to show response N. This 'pessimistic' judgement bias compares with negative cognitive biases observed in people in negative mood states. Variants on this procedure include making training cue B predictive of a negative event (e.g. noise) which can be avoided by performing response N (e.g. Harding *et al.*, 2004).

5.8 Animal Cognition and Animal Welfare

There is growing interest in understanding the cognitive abilities of domestic animals. Once we appreciate that they do not simply respond to their environments, to each other or to us, with a set of simple, fixed or 'unthinking' responses, we may want to rethink their place within our ethical frameworks. We may admire and appreciate their complexity just as we might admire great paintings or diverse and complex landscapes, and perhaps accord them greater respect or protection on these grounds alone.

Nevertheless, their cognitive abilities also have a more direct impact on their welfare. For example, behaviours such as cannibalism in chickens are facilitated by prior observation of companions (Cloutier *et al.*, 2002). Only by understanding the mechanisms of social learning in large flocks might it be possible to reduce the transmission rates of these harmful behaviours. In a different context, it is sometimes argued that animals cannot be trusted to make rational or considered choices in preference tests, because they are incapable of 'weighing up' the consequences. Studies of cognition can be used to investigate these claims. In one experiment, it was shown that feeding decisions were not guided solely by immediate considerations. Chickens were able to show a degree of 'self-control', selecting a strategy that meant they had to wait for a large reward rather than receiving a smaller but immediately available reward (Abeyesinghe *et al.*, 2005).

This type of evidence can give us a little more confidence in the use of preference tests to assess animal needs and desires. It also demonstrates that animals are able to measure the passage of time, and other studies have shown that, for example, pigs can learn to choose a location where they will be confined for a shorter period of time over one where confinement lasts longer (Spinka *et al.*, 1998). If animals can use cues to anticipate the duration of events, this might be used to 'reassure' them that an upcoming procedure will soon be over.

Finally, the types of cognitive abilities that domestic animals possess are relevant to the types of situations in which they might suffer. Although there is no compelling reason why *any* of the complex learning or cognitive abilities already described should be accompanied by subjective experiences or consciousness, it is usual to give domestic animals the benefit of the doubt and concede that consciousness is logically possible, or even likely. If so, then an animal that can remember past, stressful, events may have reduced welfare many days or weeks after the event itself ended. It is even possible that animals may generate 'innate' mental representations of things they have never encountered in their own lifetime, so that an animal kept in social isolation could still, conceivably, 'miss' the company of others. And an animal that can anticipate future events may be able to anticipate its own future. Much work remains to be done.

References

Abeyesinghe, S.M., Nicol, C.J., Hartnell, S.J. and Wathes, C.M. (2005) Can domestic fowl show self-control? *Animal Behaviour* 70, 1–11.

Bailey, D.W. and Sims, P.L. (1998) Association of food quality and locations by cattle. *Journal of Range Management* 51, 2–8.

Boissy, A., Arnould, C., Chaillou, E., Desire, L., Duvaux-Ponter, C., Greiveldinger, L., Leterrier, C., Richard, S., Roussel, S., Saint-Dizier, H., Meunier-Salaun, M.C., Valance, D. and

Veissier, I. (2007) Emotions and cognition: new approach to animal welfare. *Animal Welfare* 16(S), 37–43.

Cloutier, S., Newberry, R.C., Honda, K. and Alldredge, J.R. (2002) Cannibalistic behaviour spread by social learning. *Animal Behaviour* 63, 1153–1162.

Cooper. J.J., Ashton, C., Bishop, S., West, R., Mills, D.S. and Young, R.J. (2003) Clever hounds: social cognition in the domestic dog (*Canis familiaris*). *Applied Animal Behaviour Science* 81, 229–244.

Dickinson, A. (1980) *Contemporary Animal Learning Theory*. Cambridge University Press, Cambridge, UK.

Dumont, B. and Petit, M. (1998) Spatial memory of sheep at pasture. *Applied Animal Behaviour Science* 60, 45–53.

Fiset, S., Beaulieu, C., Le Blanc, V. and Dube, L. (2007) Spatial memory of domestic dogs (*Canis familiaris*) for hidden objects in a detour task. *Journal of Experimental Psychology: Animal Behavior Processes* 33, 497–508.

Forkman, B. (2001) Domestic hens have declarative representations. *Animal Cognition* 3, 135–137.

Freire, R., Cheng, H.W. and Nicol, C.J. (2004). Development of spatial memory in occlusion-experienced domestic chicks. *Animal Behaviour* 67, 141–150.

Hall, G. (1994) Pavlovian conditioning. In: Mackintosh, N.J. (ed.) *Animal Learning and Cognition*. Academic Press, San Diego, California, pp. 15–43.

Hanggi, E.B. (1999) Categorization learning in horses (*Equus caballus*). *Journal of Comparative Psychology* 113, 243–252.

Harding, E.J., Paul, E.S. and Mendl, M. (2004) Cognitive bias and affective state. *Nature* 427, 312.

Held, S., Mendl, M., Devereux, C. and Byrne, R.W. (2001) Behaviour of domestic pigs in a visual perspective taking task. *Behaviour* 138, 1337–1354.

Held, S., Mendl, M., Devereux, C. and Byrne, R.W. (2002) Foraging pigs alter their behaviour in response to exploitation. *Animal Behaviour* 64, 157–166.

Hothersall, B., Harris, P.A. and Nicol, C.J. (2009) Foals preferentially utilise relative spatial cues over visual cues in a discrimination learning task. *Animal Cognition.* In press.

Kendrick, K.M., da Costa, A.P., Leigh, A.E., Hinton, M.R. and Peirce, J.W. (2007) Sheep don't forget a face. *Nature* 447, 346.

Lea, S.E.G., Wills, A.J. and Ryan, C.M.E. (2006) Why are artificial polymorphous concepts so hard for birds to learn? *Quarterly Journal of Experimental Psychology* 59, 251–267.

Manteuffel, G., Puppe, B. and Schon, P.C. (2004) Vocalisation of farm animals as a measure of welfare. *Applied Animal Behaviour Science* 88, 163–182.

Mendl, M., Laughlin, K. and Hitchcock, D. (1997) Pigs in space: spatial memory and its susceptibility to interference. *Animal Behaviour* 54, 1491–1508.

Mendl, M., Burman, O., Laughlin, K. and Paul, E. (2001) Animal memory and animal welfare. *Animal Welfare* 10, S141–S159.

Nicol, C.J. (2004) Development, direction and damage limitation: social learning in domestic fowl. *Learning and Behavior* 32, 72–81.

Palleroni, A., Hauser, M. and Marler, P. (2005) Do responses of galliform birds vary adaptively with predator size? *Animal Cognition* 8, 200–210.

Paul, E.S., Harding, E.J. and Mendl, M. (2005) Measuring emotional processes in animals: the utility of a cognitive approach. *Neuroscience and Biobehavioral Reviews* 29, 469–491.

Pepperberg, I.M. (2002) The value of the Piagetian framework for comparative cognitive studies. *Animal Cognition* 5, 177–182.

Perrin, G., Ferreira, G., Meurisse, M., Verdin, S., Mouly, A.M. and Levy, F. (2007) Social recognition memory requires protein synthesis after reactivation. *Behavioral Neuroscience* 121, 148–155.

Regolin, L., Marconato, F. and Vallortigara, G. (2004) Hemispheric differences in the recognition of partly occluded objects by newly hatched domestic chicks (*Gallus gallus*). *Animal Cognition* 7, 162–170.

Shettleworth, S. (1998) *Cognition, Evolution, and Behavior*. Oxford University Press, Oxford, UK.

Spinka, M., Duncan, I.J.H. and Widowski, T.M. (1998) Do pigs prefer short-term to medium-term confinement? *Applied Animal Behaviour Science* 58, 221–232.

Whiten, A., Horner, I., Litchfield, C.A. and Marshall-Pescini, S. (2004) How do apes ape? *Learning and Behavior* 32, 36–52.

Williams, J.L., Friend, T.H., Nevill, C.H. and Archer, G. (2004) The efficacy of a secondary reinforcer (clicker) during acquisition and extinction of an operant task in horses. *Applied Animal Behaviour Science* 88, 331–341.

M. Mendl and C.J. Nicol

Social and Reproductive Behaviour

D.M. Weary and D. Fraser

6.1 What is Social Behaviour?

Social behaviour is the glue that allows groups of animals to function and that allows us to interact with the domestic animals we care for. Social behaviour happens whenever animals interact: crows fighting over picnic scraps, two people kissing at a bus stop, a dairy cow licking her newborn calf or dogs wrestling in the park. In this chapter, we provide an introduction to the basics of social interactions among animals, and we review some of the main types of social behaviour: competition, sexual behaviour, parent–offspring interactions and play. Let's begin by considering three basic questions: (i) how do animals benefit from living in groups; (ii) how do they communicate with each other; and (iii) how did these social behaviours evolve?

6.2 Group Living

Domestic animals are typically kept in groups. Indeed, given that essentially all domestic species are descended from group-living ancestors, the ability to live in groups seems to have been a prerequisite for domestication (Price, 2002). Animals in the wild may benefit in several ways from forming groups. If all else is equal, a member of a pair should be half as likely to be eaten by a predator as would the same animal on its own. In this way animals may be able to dilute their risk of predation by forming simple aggregations (Treves, 2000), although this benefit will be reduced if larger groups are more likely to be detected. Risk dilution may also be shared unequally among group members. For example, animals may benefit more from group membership if they are at the centre of the herd rather than on the periphery, or by associating with more vulnerable herd mates. According to Canadian folk wisdom, it is best to travel in bear country with companions you can outrun.

Groups may also be of value in the defence of vulnerable young. Predators will sometimes attack the young of large animals such as musk oxen or buffalo, but a group of adults acting in concert can create a formidable deterrent.

Group membership may also help in detecting danger. In particular, large groups have many eyes, ears and noses for early detection of predators. Thus groups may be able to detect danger more reliably or more quickly than solitary individuals. Also, frequent scanning for predators reduces the efficiency of other behaviour such as foraging. By forming an association with others that 'share' the cost of scanning, each group member may be able to forage more efficiently.

An obvious cost of foraging in a group is that animals may have to share the food that they discover (Giraldeau and Caraco, 2000). This cost can be exploited by some individuals through the adoption of distinct strategies or combinations of strategies: some may be food finders while others are 'scroungers' that simply exploit the food. For example, a foraging bald eagle can look for salmon on its own, or simply wait

---

for another eagle to find the fish and then steal or share in the meal. The choice of strategy will depend partly on the proportion of individuals using the two strategies (scroungers will do well when they are few) and partly on any inherent advantage of being the finder (such as gaining longer access).

Group foraging can also lead to more efficient detection of food, especially when food sources are distributed in patches with relatively barren areas in between. This benefit will reduce the variation in intake over a fixed period of time, a factor that may be particularly important to animals such as small birds that face the risk of starvation overnight if they do not find food on a given winter's day. Feeding in groups can also improve the probability of capturing and consuming prey. For example, packs of wolves can kill moose that no single wolf could manage, and groups of orcas can surround and capture schools of salmon.

Finally, animals in groups can benefit from learning from each other (Nicol, 2006). For example, by observing hens and older social partners, chicks can learn which food items are most profitable and which are harmful.

6.3 Communication

Almost all social behaviour involves some form of communication (Hauser, 1996). When a young calf becomes separated from the cow it will call repeatedly. When the cow hears these calls it will turn toward the sound, attempt to approach the caller and vocalize. Thus, signallers can affect the behaviour of receivers by the signals they produce. Communication can occur through a range of modalities – by sound as in the current example, but also by smell (or other chemical detection), sight and touch.

Humans, having a poorly developed sense of smell, find it hard to appreciate the importance of smell for many of the domestic mammals. Dogs, for example, are 100 million times more sensitive than humans to some chemicals. Many species use chemical signals to communicate such things as territorial boundaries and reproductive status. Scents have the advantage that they can be deposited and serve as a marker long after the signaller has left. Some mammals also possess a vomeronasal organ, located at the base of their nasal cavity, which is used to detect chemical signals; air is passed over this organ when animals show 'flehmen' or lip curling, such as when a stallion detects a mare in heat.

When animals are in relatively close range or live in an open habitat they can use visual signals such as antlers on deer, colourful plumage on some birds or body movements such as holding the tail erect by dominant dogs. Body movements have the advantage of being flexible and hence well suited for signalling information that is likely to change depending on the signaller's condition. More static displays, such as plumage coloration, are useful for signalling more stable information such an animal's species, sex or individual identity.

Tactile signals include social grooming in primates, suckling and nuzzling in mammals and many of the behaviours associated with social play. One complex example is the nuzzling of the sow's udder by her piglets during a nursing bout. Litters signal their desire to nurse by gathering at the udder and nuzzling it with their snouts, even though no milk is available at that time. In response to the nuzzling, the sow will often lie on her side, rotate her udder so that the teats are exposed and grunt rhythmically. If enough piglets are present and continue massaging the teats, the sow

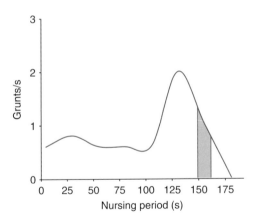

Fig. 6.1. An example of a nursing bout from a domestic pig, showing the relationship between the sow's grunt rate and milk ejection. At the beginning of the episode, sows grunt rhythmically as the piglets massage the udder with their snouts. The rate of grunting then increases rapidly at or near the time when oxytocin is released from the pituitary gland into the bloodstream. About 20 s later, the oxytocin reaches the udder and causes an ejection of milk (shaded area). The piglets generally continue to massage the udder during the slow grunts and then switch to sucking on the teats, apparently in anticipation of the milk, when the rate of grunting increases (adapted from Whittemore and Fraser, 1974).

will probably respond by releasing oxytocin from the pituitary gland. The hormone then travels through the bloodstream to the udder where it triggers an ejection of milk. However, if the nuzzling stimulus is too weak, for example because some of the piglets have not assembled, then the release of oxytocin will not occur and the piglets will have to try again later. This prevents a few piglets from triggering a milk ejection (and making pigs of themselves!) when the rest of the litter is asleep. Interestingly, the sow increases her rate of grunting at the time of oxytocin release (see Fig. 6.1). When the piglets hear this vocal signal, they stop their vigorous nuzzling of the udder and switch to sucking on the teats in anticipation of the arrival of milk.

The relative cost and benefit of communication may help to explain why animals communicate in some situations but not in others. For example, some animals vocalize when they are in pain whereas other are stoic. The difference in behaviour probably reflects differences in the potential audience. Dependent young, such as newborn piglets, may benefit by attracting a parent if they signal pain. In contrast, an adult dairy cow may gain little from signalling pain associated with lameness. Indeed, in a natural setting, signalling such pain might alert predators to the cow's higher vulnerability. Thus, we think of communication, like other types of behaviour, as having been shaped by natural selection to maximize the animals' chances of survival and reproduction.

6.4 Natural Selection

How did social behaviour evolve? To answer this question, it helps to divide social behaviour into three broad categories:

- Mutualistic behaviour (+/+): where both the 'actor' and the 'recipients' of the behaviour benefit.
- Selfish behaviour (+/–): where the actor benefits but recipients experience a cost.
- Altruistic behaviour (–/+): where the actor experiences a cost while recipients benefit.

The evolution of mutualistic behaviour comes as no surprise. We would expect natural selection to favour types of social behaviour which, on average, benefit all the animals that show it. Thus, for example, we might expect that fish would evolve the behaviour of schooling, assuming that all the animals are typically better off for behaving in this way.

Selfish behaviour is also easy to explain. Because natural selection operates powerfully at the level of the individual, we expect behavioural traits to evolve if they increase the reproductive success of the individuals possessing them, even if they have negative consequences for other members of their groups. For example, when a vulture joins other vultures at a carcass, it obviously benefits from access to food even if this means that other group members get less.

Altruistic behaviour, however, seems puzzling. In altruistic interactions individuals behave in ways that are costly to themselves but offer benefits to group-mates. How do such behaviours persist in populations when they reduce the reproductive success of the individual performing them? There are, in fact, two common ways in which these behaviours can be selected for: through kin selection and through reciprocation.

In kin selection, heritable altruistic behaviours persist when they benefit an individual's relatives. The genes that code for these behaviours increase in the population because they increase the reproductive success of these relatives rather than of the individual performing the behaviour. The evolutionary biologist W.D. Hamilton summarized this selection process in a simple formula (see Box 6.1). He posited that kin-selected altruism should occur where the reproductive cost of an act to the individual is less than the reproductive benefit to the recipient divided by the probability that the recipient also carries the genes for the altruistic trait. Full siblings share half of their genes, thus the probability of one sibling carrying an altruistic trait exhibited by the other is 0.5 ($r = 0.5$). As the famous biologist J.B.S. Haldane explained, we should be willing to lay down our lives for two full siblings or eight first cousins!

Kin-selected altruism is common in species that form groups of closely related individuals (e.g. bees, wasps or ants). Worker honey bees from the same patriline, for example, share 75% of their genes with their sisters. These workers spend their lives helping to rear the queen's daughters and do not, themselves, reproduce. This behaviour can be understood in part because of the benefits derived from kin selection: because the workers are closely related to the queen's offspring, their behaviour increases the frequency of these altruistic genes via the fertile offspring of the queen.

Altruism may occur between non-relatives under conditions that allow for reciprocation. In 'reciprocal altruism' an individual will incur a cost to help others but will later benefit when these individuals come to their aid. Because cheaters can easily exploit this behaviour, we would expect it to occur only where social networks are stable enough for individuals to encounter each other frequently and where individuals can identify and punish cheaters.

D.M. Weary and D. Fraser

Box 6.1. Hamilton's Rule

B(r) > C

Kin-selected altruism should occur where the reproductive cost (C) of an act to the individual is less than the reproductive benefit (B) to the recipient divided by the probability that the recipient also carries the genes for the altruistic trait (called the coefficient of relatedness between the two individuals (r)).

In the drawing shown, a 1-year-old scrub jay (*Aphelocoma coerulescens*) feeds his parents' current offspring, a common behaviour in the species. The gene for this altruistic 'helping' behaviour is represented as the shaded portion of the circles imposed on each bird. The juvenile helper inherited the trait from one parent (upper left) and, as the nestlings have the same parents, the probability of each nestling carrying this trait is 0.5 (r). For this behaviour to persist in the population, the benefit must be more than double the cost. Expressed mathematically, B(0.5) > C or B > 2C.

Some of the most interesting instances of altruistic behaviour appear to be due to natural selection acting at the level of the group rather than the individual. In these cases individuals pay some cost, such as reduced reproduction, to the benefit of the group. The conditions under which such 'group selection' can occur in nature are limited, but artificial selection in domestic animals can be arranged so as to favour altruistic behaviour among group members. If poultry geneticists select individual birds with the highest egg production in a group, these may breed inadvertently for aggressive, competitive behaviour. If, however, they keep closely related birds together in groups, and then breed selectively from the groups that achieve high average production, then they will breed for an ability to do well in a social setting. Given that laying hens are typically housed in groups on commercial egg farms, the latter would seem to be a better strategy. Indeed, experimental work has shown that such group-level selection can lead, in just a few generations, to birds that are both highly productive and relatively non-aggressive (Muir and Craig, 1998). This is a promising area for new research on other domestic animals we house in groups.

6.5 Common Social Behaviours

Competition

Why do some animals defend resources from competitors whereas others simply use resources without trying to defend them? Imagine a group of sows foraging for worms in a freshly ploughed field. The worms will be distributed fairly evenly in

---

space, so an animal would gain little from defending any specific part of the field – in fact, the time spent in defence would only be lost from foraging. The sows may still be engaged in a kind of competition, in the sense that the animals that find and eat the worms most efficiently leave fewer for the others. However, this so-called 'scramble competition' is indirect and not overtly aggressive. Now imagine the same group of sows fed a concentrated diet from a single feeder. In this situation, an aggressive sow may well benefit from defending the food source from other sows ('territorial defence'), because the time and effort needed to exclude competitors may allow the territory holder to consume extra food.

Which of these two patterns we witness will depend partly on how resources are distributed in both time and space. As we see in this example, it is generally less beneficial to defend resources that are spread over a large area. How resources are distributed over time will also influence the type of competition. For example, feed delivered gradually over the day as a slow trickle is more easily defended than an equivalent quantity delivered all at once.

When resources are defended, we expect the outcome of contests to be decided by differences in factors such as body size, strength or competitive abilities – factors collectively known as 'resource-holding potential' or RHP (Pusey and Packer, 1997). Animals that compete for resources need to become skilled at assessing each other's RHP so as to avoid risking injury and wasting effort by engaging in fights that are likely to be lost. Newly mixed pigs spend considerable time fighting, and these fights can lead to injuries. However, these fights are longer and more serious when animals are similar in RHP. Mixing animals that differ substantially in body size can help reduce the amount of fighting.

Sometimes other sorts of asymmetries can affect which animal wins a contest. One of the best known relates to the advantage of being the current resource holder. Territory holders may benefit from better knowledge of the territory. Also, intruders must often compete not only with the resident but also with the neighbouring territory holders – a cost that an established resident does not have to pay. Animals can also vary in their need for a specific resource, and this too can affect competitive behaviour. For example, a lioness may defend a portion of a carcass for a period of time but, as she becomes satiated, she is more likely to be usurped by hungry pride-mates.

When individual animals are in frequent competition over resources, they can avoid the cost of continual aggression by establishing which one is 'boss'. In the 1920s an ethologist studying small flocks of chickens noticed that one bird in each flock tended to peck all the others, whereas a second bird pecked all but the first, a third pecked all but the first two, and so on. This simple relationship came to be called the 'peck order' or 'dominance order' of the flock (see Fig. 6.2a). This idea took wing (!), and soon animal behaviourists were describing social behaviour in other situations, from rat colonies to office politics, in terms of dominance orders, often by testing animals to determine which would have priority of access to resources such as food.

In reality, social relationships between animals are generally far more varied and complex than a simple dominance order would suggest (Langbein and Puppe, 2004). For many species, although it may be possible to rank the animals based on their access to resources, a simple hierarchy may misrepresent the kind of social behaviour seen in the group. In long-established groups of pigs, for example, the animals often

D.M. Weary and D. Fraser

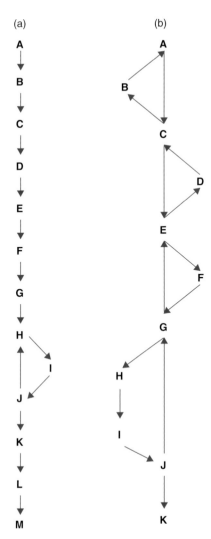

Fig. 6.2. Classic examples of the 'peck order' in two small flocks of chickens. Birds are identified by letters corresponding to 'rank', where bird A pecked the greatest number of other birds, B pecked the second greatest number, and so on. Each bird pecked all birds below it in the diagram, except as shown by arrows pointing upward. (a) Thirteen young females showed an almost perfectly linear order except that bird J pecked the higher-ranking H; (b) ten young males showed a complex set of relations, with more than half of the birds pecking higher-ranking flock-mates (data from Masure and Allee, 1934).

show mild mutual aggression with no clear hierarchy. Other species have multiple forms of social relationships: in cattle, for instance, individuals with high priority of access to food may not be the ones that normally lead the herd to new locations. Even dominance relations among chickens can depart greatly from any simple pecking order (see Fig. 6.2b). A focus on dominance also tends to emphasize competition between animals, whereas some relationships are characterized more by cooperation and mutual tolerance.

Sexual behaviour

When people breed animals on farms or in zoos, they generally select which males will mate with which females in order to achieve the desired combinations of genes in the offspring, and then hope that the chosen parents will cooperate by mating. In the wild, however, the animals themselves make these choices. The scientific study of mate selection has begun to uncover the underlying principles.

The importance of mate choice to males and females depends on the time, energy and other resources that the animals invest in producing the offspring (Clutton-Brock, 1991). Some animals (particularly the males in many mammalian species) contribute only gametes (sperm), which are relatively inexpensive to produce, whereas other animals (mammalian females) incur much larger reproductive costs such as maintaining pregnancy and feeding the young by producing milk. In general, we expect that the sex that makes the larger investment (and thus has a lower potential reproductive rate) will be the most selective. In all of the common domestic animals, the male is able to father young at a much faster rate than females can produce them, so we expect females to be selective in their choice of mates. In the wild, the elaborate breeding displays and feather coloration of males in many bird species – traits that are attractive to females but may also increase the birds' vulnerability to predation – illustrate the strong selective pressures that have shaped males to becoming attractive to females.

Mating systems can be divided roughly into four categories: monogamy, polygyny, polyandry and promiscuity. In monogamous systems male and female bond for some period of time and often both parents contribute to care of the offspring. Polygynous systems are characterized by males mating with multiple females and females mating only with a single male and normally caring for the young. Polyandrous systems, seen especially in some species of fish, are the reverse. Promiscuous systems involve a mixture of polygyny and polyandry. Although the ancestors of domestic species may have employed a different mating system, domestic animals are typically promiscuous or polygynous. For example, whereas wolves are monogamous, domestic dogs tend to be promiscuous. The seasonality of breeding is also reduced in domestic animals, with most domestic species able to breed year-round.

Polygynous males can gain mating opportunities by controlling access to resources, such as food or nesting sites, that are important for female reproduction. This 'resource defence polygyny' is facilitated by factors that help make resources defendable, such as the clumping of suitable resources in space. Males that control key resources will be more likely to mate or find multiple mates. For example, male red-winged blackbirds that have territories containing the most suitable nesting sites are more likely to attract multiple females as mates. Alternatively, polygynous males can attempt to defend a group of females as a harem. This is facilitated when females group for other reasons such as for anti-predation benefits, as discussed above. Among red deer, for example, stags compete with other males for control of harems, and somewhat similar competition among males is common in some domestic species.

Parent–offspring interactions

Some of the most important social interactions occur between parents and their dependent young. Parents typically prepare suitable locations for the young, provide them with nutrients and protect them from harm, although how these goals are achieved varies enormously from species to species.

Pigs and sheep provide an interesting contrast. The sow gives birth to a large number of small, fragile piglets. Typically she separates herself from her usual social group a day or so before farrowing, and seeks a secluded nest site. She prepares the site by rooting the soil to form a soft depression, and then lines it with resilient material such as branches, and with soft, insulating material such as grass. While the piglets are being born she lies relatively immobile with the udder exposed, leaving the piglets to find the teats unassisted. The carefully prepared nest site protects the piglets from cold and keeps the young in a single location that the mother can protect. With her relatively passive behaviour during parturition, the sow avoids harming the piglets through excessive movement, but does not learn to discriminate between her own and foreign young for perhaps a day or more. Hence, it is relatively easy to foster day-old piglets, from a sow that has too many young to one with a smaller litter.

Ewes, which normally give birth to only one or two large, mobile offspring, follow a different pattern. A ewe may move away from other sheep for parturition, or give birth in the midst of the flock. Once a lamb is born, the ewe licks it vigorously; this stimulates the lamb to rise and suckle, and it exposes the mother to the odour of the lamb, such that she can tell her lamb from others soon after birth. Moreover, lambs are exposed to the mother's voice while still in the uterus, and they seem to respond specifically to the calls of their own mother, more or less from birth. Thus, although the lamb and ewe are highly mobile, they remain close together in the flock through mutual recognition and attraction. If farmers want to foster an orphan lamb on to a ewe whose own lamb has died, they often have to go to great lengths to disguise the appearance and odour of the orphan or the ewe will refuse to let it suckle.

Given its importance to fitness, parental behaviour is underlain by strong motivations. Under the influence of hormonal changes before parturition, the sow becomes extremely restless and, even if confined in a narrow stall, will still greatly increase her level of activity, standing and lying repeatedly and showing a strong interest in nesting material. Once the young are born, the sow may become protective if intruders approach the nest, and, if a piglet escapes from the nest and gives characteristic 'separation calls', the sow becomes very attentive and seeks out the lost animal. Maternal motivation in the ewe takes a somewhat different form. Ewes that are about to lamb often become intensely attracted to newborn lambs, and may even 'steal' newborns from other ewes, but this behaviour ceases once her own lamb is born.

Traditionally, many animal behaviourists studied parent–offspring behaviour as if the relationship were all one-way, with the parent providing care for relatively passive young. In reality, the young have signals – sometimes subtle ones – that solicit care from the parent. Many young mammals and birds have some form of distress call when separated, and hunger signals that encourage the parents to provide more food. In a famous demonstration, ethologist Lars von Haartman showed the power of these begging calls. He studied pied flycatchers that were raising their young in specially designed nest boxes containing a hidden compartment where von Haartman could temporarily conceal a second brood of hungry nestlings. Exposed to a double dose of begging calls, from both their own brood and the concealed brood, the parents increased their rate of bringing food to the nest, to the benefit of their fortunate offspring.

Given the ability of the young to stimulate parental care, parents need to strike a balance between providing too much care and too little. If parents devote too many resources to their present young, they may be slow to re-breed or not be in good enough condition to raise their next young successfully. The young, however, having a greater stake in their own success than in that of future siblings, may solicit a higher level of care than the parents ought to provide for their own maximum reproductive success. One obvious outcome of these conflicting interests is 'weaning conflict', whereby the parent may repel attempts by the young to suckle or solicit food. Some housing conditions make it difficult for the mother to escape the offspring's demands. For example, if sows are able to get away from their piglets, they gradually reduce the frequency of nursing and thus force the piglets to eat more solid food but, if the sow and litter are kept permanently together, the sow fails to reduce the frequency of nursing and may lose much more body condition through the heavy demands of lactation (see Fig. 6.3; Drake *et al.*, 2008).

Play

One of the most interesting but under-researched aspects of social behaviour is play (Spinka *et al.*, 2001). Social play often consists of interactions that imitate the process, if not the end point, of more clearly functional behaviours like fighting and hunting. Sometimes animals use specific behaviours that signal the onset of play. Domestic dogs, for example, will often 'bow' during play sequences, especially if play includes fighting behaviours that could trigger aggression from the playmate.

How animals benefit from play is not well known. Play may serve as practice for adult activities, allow animals to become familiar with their environment and develop social skills and relationships. Social skills are especially important for species that need practice to develop effective courtship, appeasement or competitive behaviour.

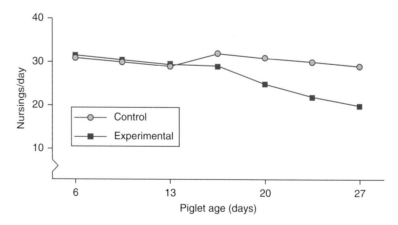

Fig. 6.3. The number of nursing episodes per day by sows in relation to the age of the piglets. Sows confined with their piglets (control, circles) maintained a similar number of nursings per day throughout the 27 days of lactation. Those that could leave the young by stepping over a low barrier (experimental, squares) gradually reduced the number of nursings, starting in the third week of lactation (adapted from Weary *et al.*, 2002).

D.M. Weary and D. Fraser

In a series of experiments that were both famous and infamous, psychologist Harry Harlow raised infant rhesus macaques in individual cages where they had no physical contact with either their mothers or other young monkeys. The objective was to raise very healthy animals by keeping them free from infectious diseases, but Harlow soon discovered that the animals showed signs of emotional disturbance and dysfunctional social behaviour: they stared into space, circled in their cages in a stereotyped manner and rocked repetitively. However, Harlow found that as little as 20 min/day of play with other monkeys in an enriched environment was enough for reasonably normal social behaviour to develop.

In the farmed animals play is likely less critical than it is for primates, but it may still serve a valuable role. For example, pre-weaned dairy calves that are reared in groups (as compared with the more conventional individual housing) spend time playing, and these calves are more likely to become dominant when mixed with animals that have been individually reared. Play may also prepare animals for coping with unusual situations, such as maintaining balance on a slippery surface. Indeed, play sequences often involve some aspect of self-handicap, such as adopting ungainly postures, and this may allow young animals to develop physical skills that will prove useful later in life.

6.6 Conclusions

For millennia the domestication of animals has depended on the ability of humans to understand the social behaviour of animals, to accommodate this behaviour in controlled circumstances and sometimes to become part of the animals' social world. There is great scope for improving our care and handling of domestic animals by further improving our understanding of their social behaviour.

Most of the theories, concepts and vocabulary described in this chapter have been developed by behavioural scientists to help us understand how behaviour equips animals to live in the environment in which the species evolved. With most domestic animals, many of the behavioural adaptations of the species are still present, but may be ill suited to the unnatural physical and social environment in which the animals are kept. Like its wild counterpart, the newly hatched domestic chick appears predisposed to become imprinted on a parental figure, but what form does imprinting take when hundreds of chicks are hatched together in an incubator? Like the wild sow, the domestic sow seems predisposed to wean her young gradually by spending less and less time with them as they age, but, what are the consequences if the sow and litter are penned together continuously for several weeks, and then abruptly and permanently separated? Females of many species seem predisposed to select mates based on certain attributes, but, if they are penned with only a single male, does this affect their willingness to mate or how strongly they display oestrus? One area of special interest is how social behaviours, which probably evolved to facilitate interactions among small groups of related individuals, can break down when animals are housed in pens containing hundreds of unrelated animals (Croney and Newberry, 2007). Thus, a key challenge for future research is to understand how rearing conditions can be altered better to suit the social behaviour of animals, and thus help avoid problems for both the animals and the people that work with them.

References

Clutton-Brock, T.H. (1991) *The Evolution of Parental Care*. Princeton University Press, Princeton, New Jersey, 352 pp.

Croney, C.C. and Newberry, R.C. (2007). Group size and cognitive processes. *Applied Animal Behaviour Science* 103, 215–228.

Drake, A., Fraser, D. and Weary, D.M. (2008) Parent–offspring resource allocation in domestic pigs. *Behavioral Ecology and Sociobiology* 62, 309–319.

Giraldeau, L.A. and Caraco, T. (2000) *Social Foraging Theory*. Princeton University Press, Princeton, New Jersey, 362 pp.

Hauser, M.D. (1996) *The Evolution of Communication*. MIT Press, Cambridge, Massachusetts, 760 pp.

Langbein, J. and Puppe, B. (2004). Analysing dominance relationships by sociometric methods – a plea for a more standardised and precise approach in farm animals. *Applied Animal Behaviour Science* 87, 293–315.

Masure, R.H. and Allee, W.C. (1934) The social order in flocks of the common chicken and pigeon. *Auk* 51, 306–327.

Muir, W.M. and Craig, J.V. (1998) Improving animal well-being through genetic selection. *Poultry Science* 77, 1781–1788.

Nicol, C. (2006) How animals learn from each other. *Applied Animal Behaviour Science* 100, 58–63.

Price, E.O. (2002) *Animal Domestication and Behavior*. CAB International, Wallingford, UK.

Pusey, A.E. and Packer, C. (1997) The ecology of relationships. In: Krebs, J.R. and Davies, N.B. (eds) *Behavioural Ecology: an Evolutionary Approach*. Blackwell, Oxford, UK, pp. 254–283.

Spinka, M., Newberry, R.C. and Bekoff, M. (2001) Mammalian play: training for the unexpected. *Quarterly Review of Biology* 76, 141–168.

Treves, A. (2000) Theory and methods in studies of vigilance and aggregation. *Animal Behaviour* 60, 711–722.

Weary, D.M., Pajor, E.A., Bonenfant, M., Fraser, D. and Kramer, D.L. (2002) Alternative housing for sows and litters. Part 4. Effects of sow-controlled housing combined with a communal piglet area on pre- and post-weaning behaviour and performance. *Applied Animal Behaviour Science* 76, 279–290.

Whittemore, C.T. and Fraser, D. (1974) The nursing and suckling behaviour of pigs. II. Vocalization of the sow in relation to suckling behaviour and milk ejection. *British Veterinary Journal* 130, 346–356.

7 Abnormal Behaviour, Stress and Welfare

L. KEELING AND P. JENSEN

7.1 Introduction

When man first started keeping animals in captivity, his concerns were probably restricted to how the animals could be prevented from escaping and how they could be kept alive and healthy. Later, we started to be concerned about production and how we could get our farm animals to produce more milk and eggs, grow faster and have more offspring or, in the case of our sport and companion animals, run faster and look more beautiful. During these times behavioural disturbances and stress were a problem only in so far as they affected health and performance, and good health was usually considered synonymous with good welfare. It wasn't until the 1960s and 1970s that we started to question these assumptions and to consider behaviour as an important component of welfare. This chapter deals with the important aspects of behavioural disorders and stress and ends with a short review of current theories on what animal welfare is and how to measure it. Other books taking up this subject in more detail include Broom and Johnson (1993), Appleby and Hughes (1997) and Broom and Fraser (2007).

7.2 Behavioural Disorders

Normal and abnormal behaviour

One way to define normal – or natural – behaviour of an animal could be the behaviour that has developed during evolutionary adaptation. This would, of course, include any learnt behaviour that serves the function of promoting the health, survival and reproduction of an animal in a certain environment. For domesticated animals, there are three important sources of information about normal behaviour: (i) the behaviour of the wild ancestors; (ii) the behaviour of feral animals, i.e. domestic animals that have escaped or been released and have adapted to a life without dependence on humans; and (iii) the behaviour of domestic animals when placed (usually by researchers) in environments similar to those of their ancestors (see Fig. 7.1).

Since learning and adaptation will modify the behaviour of any individual, there will always be a range of behavioural profiles that can be considered normal. Sometimes, therefore, people dismiss the very concept of normal behaviour, and consider any deviations to be adaptive. However, in spite of a large variation, there are aspects of the behaviour of any animal that are typical for the species, and there may be aspects which are outside the range usually observed in members of the species in non-captive situations. It is therefore necessary to try to understand which behaviour patterns are species-typical and which ones can be regarded as not normal for an animal of a given species, sex and age.

Fig. 7.1. When pigs are released back into nature they will form groups of closely related sows, as do wild boars, and spend most of their time in pursuit of food, mainly by rooting.

To avoid semantic confusion, it is important to remember that, in this context, an abnormal behaviour may be very common among individuals of a species. For example, since most poultry in the world are kept in captivity, and most of the egg-laying hens are kept in cages, any abnormal behaviour caused by the cage environment is going to be shown by a majority of the hens. Frequency of occurrence, therefore, is no synonym for normality with respect to behaviour. When we consider abnormal behaviour and behavioural disturbances, we must always remember that the norm is the behaviour as it has evolved in the natural habitats of the species.

Stereotypies

Stereotypies are one particular form of abnormal behaviour. They can be described as movements that are repeated in the same manner over and over again and may occupy a substantial part of the time for which an animal is awake. They are sometimes conspicuous, but sometimes less obvious to a casual observer. Stereotypies have been defined as movement patterns that are unvarying, repetitive and lack an obvious goal or function (see Mason, 1991, for a review, and Chapter 4, this volume, for some further details on this).

Stereotypies are thought to develop when an animal is prevented from performing certain normal and strongly motivated behaviour patterns, for example those related to foraging and exploration. A restrictive and stimulus-poor environment has been shown to affect the probability of an animal developing stereotyped behaviour,

L. Keeling and P. Jensen

although in some cases the crucial factor appears to be whether or not the animal has the opportunity to forage. For example, experiments with tethered and loose-housed sows found that animals on a restricted diet developed more stereotyped behaviour patterns, such as sham chewing, bar biting and chain chewing, than sows on the higher ration, irrespective of housing condition (see Fig. 7.2). Differences between animals in the way they eat are reflected in the range of stereotypies that are seen, e.g. chewing stereotypies in pigs, tongue-rolling stereotypies in cattle, pecking stereotypies in chickens and crib biting in horses. In general, it appears that grazing and omnivorous animals, which are likely to be motivated to spend much time feeding, tend mainly to develop oral stereotypies (e.g. biting, chewing, pecking), whereas predators, which are likely to be motivated for moving and chasing, more often develop locomotory stereotypies (e.g. pacing, route-tracing).

Efforts to investigate the link between the performance of stereotypic behaviour and welfare, however, have proved to be problematic, with some researchers finding a good correlation and others no relationship between stereotypies and other welfare measures. It is now thought that there are different phases in the development of stereotypies. At first the behaviour is context specific, but it is later performed in an ever-increasing range of situations until, when fully established, it continues even when the animal is moved to a more enriched environment. Stereotyped behaviour is evidence, therefore, that at some time the animal has had reduced welfare, but it does not necessarily imply that its welfare is poor now. However, stereotypies remain important indicators that the environment is not providing sufficient opportunities for the animals to perform their normal behaviour. The number of animals showing stereotypies in a particular environment and the proportion of time they spend performing stereotypies are therefore important welfare indicators.

Fig. 7.2. Sows that are confined or tethered during pregnancy spend a large part of their time performing stereotyped bar biting.

Cannibalism

Although dramatic and something most people would want to prevent, cannibalism in animals is not necessarily an abnormal behaviour. It occurs in the wild, but only under conditions of adversity, when the benefits of removing competitors from an area and the nutritional benefits of eating them outweigh the potential risks. Of the species that are dealt with in this book, cannibalism is most common in pigs, poultry and rodents. Studies in behavioural ecology have shown that, in nature, it is usually larger individuals cannibalizing smaller ones, often young, or that it is healthy individuals cannibalizing weaker ones.

However, in captivity, cannibalism may develop in poor environments, for example where the animals live in crowded or impoverished housing systems. It is a common misconception that this behaviour is a redirected form of aggression, but mostly aggressive motivation has nothing to do with cannibalism. In chickens, the most serious form of cannibalism is when pecks are directed at the cloaca, which may perhaps sometimes be a redirected investigative behaviour. There is evidence from outbreaks in commercial flocks of laying hens that it is the large birds that become cannibals and the small birds that become victims. The effects, however, are subtle and a chance peck resulting in skin damage followed by learning for the reward of blood probably also plays a role (see Fig. 7.3).

Tail biting in fattening pigs is more common when the pens are small, crowded and lacking in appropriate stimuli. Giving the pigs sufficient access to straw for exploration and chewing is a relatively good way of preventing this behaviour. It may also have some chance component, although in pigs given the opportunity to chew on blood-soaked 'imitation tails' (tassels of rope hung in the pen) nutritional deficiencies – and in particular salt deficiencies – were found to increase the incidence of chewing.

That there is some nutritional basis for cannibalism would be in accordance with outbreaks in the wild that often occur in times of food shortage. In rodents, pup cannibalism is common. In this instance cannibalism is by the parent, and can sometimes be adaptive in nature; often, the victim would have had little chance of surviving in

Fig. 7.3. Feather-pecking in laying hens is a serious welfare problem, which may sometimes develop into cannibalism.

L. Keeling and P. Jensen

any case. However, it may also be caused by stress. When performed by an unrelated individual in nature, the benefits of infanticide – i.e. eating of the young – may be represented by increased food acquisition, but there can also be decreased resource competition or increased male reproductive success.

Despite the work on cannibalism from a behavioural ecology perspective, we have only a poor understanding of all aspects of cannibalism in farm animals. Largely, it seems to be multifactorial, with factors such as bareness of the environment – e.g. lack of straw for pigs or litter for poultry, increasing stocking density and group size – all increasing the risk of cannibalism. There has been some success in reducing it by genetic selection, but whether this is a direct effect on cannibalistic behaviour or an indirect effect on some other linked trait is unclear.

Abnormal aggression

Aggression is used in the establishment of dominance relationships, and so in long-established groups the level is low and rarely a problem. Aggression is therefore a normal behaviour, but in domestic animals we regard it as undesirable if it reaches such proportions as to cause injury or practical problems. Unfortunately, it is often the way we house or manage animals that results in high levels of aggression. For example, we often mix animals or keep them in groups that would be unlikely to occur under natural conditions.

Horses, as presented in a later chapter, live in bands in the wild. However, at a typical riding stable, horses may be moved from one paddock to another on an almost daily basis and may even be part-time members, being outside only during the daytime in fine weather. Comparisons show that the level of aggression in these groups is much higher than in groups in the wild. The most common reason for outbreaks of aggression in pigs is also the mixing of unfamiliar animals. Attempts to prevent newly mixed animals from fighting have not been very successful, not even by the use of drugs.

Aggression is also influenced by stocking density, and an aggressive interaction may persist if a submissive individual is not able to signal its submission effectively because there is insufficient space for it to remove itself from the aggressor. This has been shown to be a contributing factor in aggression among fattening pigs or sows.

Finally, aggression problems in dogs are widely reported. Here, the problem may be because of the high inbuilt aggression level, but it can equally well be that the owner perceives even a low level of aggression in their pet as undesirable, for example when there are small children in the family.

7.3 Stress

The concept of stress was developed over a number of decades in the middle of the 20th century. The leading lights were two physiologists, Walter B. Cannon and Hans Selye, who both studied bodily reaction patterns to potentially harmful stimuli. The reactions appeared to be more or less the same regardless of which types of threats they studied. Their combined findings gave rise to what we might call the standard stress model (Toates, 1995).

According to this standard model, a threatening stimulus (often referred to as a 'stressor') evokes mainly two sets of physiological responses. One is the activation of the sympathetic branch of the autonomic nervous system causing, for example, increased heart rate, increased blood pressure, decreased gastrointestinal activity and increased secretion of catecholamines (adrenalin and noradrenalin) from the medulla of the adrenal cortex. Catecholamines in turn emphasize and prolong similar effects on the circulation and intestines. The other set of reactions consists of an increased secretion of the hormone ACTH (adrenocorticotropic hormone) from the pituitary (regulated by CRH, corticotropin-releasing hormone), which stimulates secretion of corticosteroids (for example, cortisol and corticosterone) from the adrenal cortex. Corticosteroids are mainly metabolic hormones, stimulating the recruitment of energy in the form of glucose and fatty acids from depots in the body (see Fig. 7.4).

Prolonged activation of this system was found to be harmful to the individual in a number of ways, causing syndromes such as stomach ulceration, cardiovascular disease and alterations in the efficiency of the immune defence system. The mechanisms involved have been found to be extremely complex and there are also several other systems involved but, basically, stress in this standard model is used as a concept describing the adverse bodily effects of certain stimuli.

Why should ethologists bother about this? There are a number of reasons. One is that the stimuli evoking the responses, the stressors, have often been described as

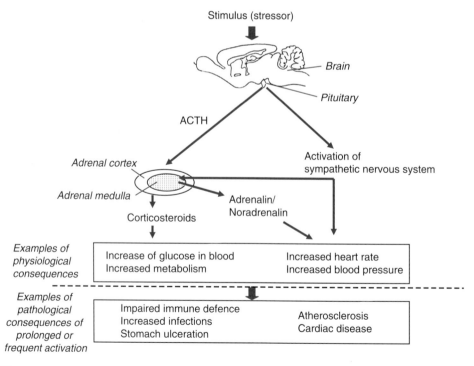

Fig. 7.4. The 'standard stress model', showing the two main physiological pathways: (i) the hypothalamic–pituitary–adrenal cortex axis (HPA axis); and (ii) the sympathetic–adrenal medulla axis (SA axis).

L. Keeling and P. Jensen

'psychological'. For example, a stressor could be to live in a crowded environment or to be exposed to sudden and unexpected novelties, both of which are of interest to applied ethologists. Another reason is that recent decades of stress research have shown that the standard model is not complete – stress responses are intimately connected with how animals perceive stressors, and to what extent they can act in a functional way in response to them. Hence, stress responses depend on the mechanisms controlling normal behaviour. Stress is therefore an important link between environment, behavioural disorders and disease (Broom and Johnson, 1993).

The concept of allostatis, which means stability through change, is increasingly being used in the context of animal welfare. This new approach suggests that changes in body functioning following exposure to a given stressor actually help the animal anticipate further challenges and so respond to them. In this way the animal can better adapt to its environment.

Predictability and controllability: key concepts in stress responses

The pathological consequences of a certain stressor have long been known to depend on how that stressor is perceived. The crucial experiment that demonstrated the salient factors in perception of threats was performed by Weiss (Weiss, 1971). He exposed rats to well-controlled stressors in the form of electrical shocks, delivered via an electrode attached to the tail. In each test, Weiss had three rats in single cages. Two of the rats were yoked in the same electric circuitry, so when one rat was shocked the other received exactly the same sensation. The third rat was a control, which never received any shocks. One of the yoked rats was then given a predictive cue, in the form of a sound signal going off a few seconds before each shock.

The possibility of predicting the shock significantly decreased the development of gastric ulceration, even though both rats had received exactly the same physical pain experience. When Weiss allowed one rat to exert some control over the shocks, the effect was even stronger. The rat could turn the current off once it had started (by turning a wheel), and thereby shorten the shock for itself and its unknown yoked partner, or postpone the shock somewhat. Again, ulceration decreased dramatically, to the extent that the rats with opportunities to predict and control had only small differences with regard to ulceration compared with those animals that never received any shock at all.

This experiment and others following have shown that the effects of a stressor do not depend so much on the physical characteristics of the stressor (intensity, duration, frequency, etc.) as on whether or not the animal can predict and, above all, control the stressor. Control is normally exerted by means of performing a behaviour relevant for the stimulus.

Individual differences in stress reactions

Even in historical stress research, it was clear that individuals often reacted quite differently to the same stressor under identical conditions. This obviously reflected some constitutional differences between individuals. In humans, psychologists would refer to 'Type A' persons for those more prone to react with activity in the sympathetic nervous system, and to 'Type B' for those with less sympathetic reactivity. Behaviourally, Type A were described as being more extrovert and aggressive, and more inclined to react with active attempts to control a stressor, while Type B would behave much in the opposite way.

This was paralleled by observations of similar differences in animals. Male tree shrews (*Tupaia belangeri*) were found to be either actively subordinate to a dominant male or passively submissive. If a submissive individual was placed in a cage close to a dominant male, it would lose weight and die within a rather short time, whereas the actively subordinate shrew would survive and live fairly well in the same situation. In wild-caught mice, resident males (males that are the 'owners' of a cage) were found to behave in one of two distinct and different ways towards an intruder: either (i) they attacked the new mouse within a very short time; or (ii) they did not attack at all. Very few intermediates, attacking for example after a few minutes, were observed, leading the researchers to conclude that there were two non-overlapping sub-populations among the original population of wild mice (Sluyter *et al.*, 1996).

A strong genetical basis for the behavioural differences between mice is shown by the fact that selection for either short or long attack latency in the resident–intruder set-up is possible. The young of short-latency parents on average attack intruders immediately, and vice versa.

Mice differ not only in their response to strangers, but in more or less any situation where they are exposed to stressful events, even very minor ones. For example, mice of both lines take the same amount of time to learn to run down a maze to reach a goal box with food. But, if a very small modification of the environment is introduced, such as placing a small piece of tape on the floor, the animals from the two populations differ in their reactions. Those selected for short attack latency seem not to pay any attention to the new item, whereas the long-attack latency mice become disrupted in their task. They seem to forget what they are supposed to do and spend their time exploring the tape instead.

The reactions of short-attack latency mice follow a general pattern, which has been termed proactive coping, whereas the pattern of the others has been called reactive coping. Proactive copers will attempt to deal with any challenge by finding routines and performing behaviour patterns that have proved to be successful previously. Reactive copers face the same challenges by attempting to modify their behaviour to find the best way of handling every new potential stressor in its own way. Physiologically, proactive copers react to challenges mainly by activation of the sympathetic nervous system, while this system is involved much less in the reactions of reactive copers (see Table 7.1).

It seems as if there is some generality to these observations across species and even phyla – there is some evidence of similar within-species differences among birds. In farm animals, the reactions of young piglets to a simple restraint test have been used to predict a variety of reactions to challenges later in life. They even have some predictive value for disease susceptibility and some aspects of meat quality following slaughter many months later (Ruis *et al.*, 2000).

Is stress a physiological or a behavioural phenomenon?

Since stress was originally discovered by physiologists and described in physiological terms, it has often been treated mainly as a physiological phenomenon. However, the

Table 7.1. A scheme showing the major differences between proactive and reactive mice.

Response	Proactive copers	Reactive copers
Reaction to social intruder	High aggression	Low aggression
Time to learn a new task, e.g. running through a maze	Same as reactive	Same as proactive
Reactions to small changes in environment	Small, pay little attention	High, investigative
Time to adapt to major changes in environment, e.g. shift in diurnal light cycle	Long	Short
Activation of HPA axis in stressful situation	Medium–high	Medium–high
Activation of SA axis in stressful situation	High	Low
Typical pathological consequences of stress	Cardiac disease, gastric ulceration	Infection, gastric ulceration

earlier described studies of Weiss, and many other similar observations, have shown that whether a certain stressor is harmful or not to an animal depends on the animal's perception of the stressor and on its possibility of behaving in a relevant way.

This has led some researchers to suggest that stress, in some aspects, can be regarded as mainly a behavioural phenomenon (Jensen and Toates, 1997). The common denominator of events that may become harmful to animals, according to this view, is that a behavioural system is motivated (see Chapter 4), but the animal cannot perform the motivationally adequate behaviour or, at least, it cannot achieve the relevant functional consequence of the behaviour. The rats in the experiment of Weiss were exposed to electric shocks in their tail, which would have motivated their flight behaviour. By performing flight motions with the forelimbs, the rats acted with the relevant behaviour (so effectively 'controlling' the situation) and, when the shock was turned off, they received the desired functional consequence. As a result, those that did not achieve the desired consequence (the ones without control) were harmed more by the stressor.

In the example above, the animal is motivated to perform a behaviour by external stimuli (in this case the electric shock), but they can also be motivated to perform a behaviour by mainly internal stimuli, such as hormones. For example, a sow due to farrow is motivated to build a nest, and this motivation is stimulated almost completely by internal hormonal events. If the sow is prevented from carrying out the nest-building activities, for example by being tethered in a crate, the increase in cortisol levels is much higher than if she is free to move around. By this measure, the sow would be said to be stressed because she is not able to act according to her motivational state.

7.4 Animal Welfare

The various discussions about abnormal behaviour in domestic animals and improved knowledge of stress have both contributed to a heightened awareness of animal welfare.

Nevertheless, in reality, it is not so simple, and not even correct, to think that good animal welfare (or animal well-being as it is sometimes called) is merely the absence of behavioural disorders and stress.

What is animal welfare?

The Brambell Committee was the first to attempt a scientific definition of animal welfare, in 1965. They were rather farsighted in three ways. First, they drew attention to the importance of behaviour in animal welfare. Up until that time good welfare had almost been synonymous with good health. Second, they stressed the importance of the scientific study of animal welfare, paving the way for future experimental studies. Third, they accepted that animals had feelings, which went against the behaviourist trend of the time (see Chapter 1). The committee also proposed five freedoms that every animal should have irrespective of how or why it is kept. These were that the animal could lie down, stand up, turn around, stretch and groom. While these may appear self-evident freedoms, there are many examples even today where they are not achievable, e.g. tethered cows and pigs which cannot turn around, or laying hens in cages which cannot flap their wings. The British Farm Animal Welfare Council later revised these five freedoms to the following:

- Freedom from thirst, hunger and malnutrition.
- Appropriate comfort and shelter.
- Prevention, or rapid diagnosis and treatment, of injury and disease.
- Freedom to display most normal patterns of behaviour.
- Freedom from fear.

Since the Brambell Committee's first attempts, many scientists have attempted to define welfare but, with hindsight, these attempts can be classified into two main categories. One category emphasizes the biological functioning of the animal (its health, reproductive success, etc.), whereas the other emphasizes the subjective experiences of the animal (suffering, pleasure, etc.).

One well-accepted definition of the first type was proposed by Broom (summarized in Broom, 1996), who defined it by saying, 'The welfare of an animal is its state as regards its attempts to cope with its environment.' Here, it is proposed that we can assess welfare by recording disease, injury, abnormal behavioural patterns and physiological changes related to stress (a high incidence of which would imply that the animal was not coping) and growth or reproduction (high rates of which would imply that the animal was coping). It is argued that, by taking many measures from behaviour, physiology and health, we get a balanced view of where the animal is on the scale from coping easily to not coping at all. One benefit of this approach is that we already have many techniques for measuring these parameters. The disadvantage is that we are not sure how to combine them in a way that takes into consideration that some measures should be given more weight than others.

One well-accepted definition of the second type was proposed by Duncan (summarized in Duncan, 1996), who stated that 'Welfare is all to do with what the animal feels.' Here, it is proposed that feelings have evolved in animals to improve survival and fitness. Pain is obviously adaptive; by making it unpleasant for an animal to put weight on an injured leg the chances of the leg healing increase. Similarly, the frantic behaviour seen in frustrating situations may increase the chances that the animal eventually changes its situation. It may even be that positive states, such as pleasure, are used for rewarding the animal for performing an appropriate behaviour, so increasing the chances that it will be shown again.

It is argued that, while under natural conditions feelings will closely reflect the health and physiology of an animal, there is a risk that for domesticated animals kept in captivity the feeling and its usual physiological correlate become separated. In view of this risk, it is argued that we must pay attention to whether the animal feels hungry or stressed, not necessarily to whether the animal has a nutritional deficit or elevated corticosteroid levels. The advantage of this approach to defining animal welfare is that it is intuitively appealing, reflecting concerns in studies related to quality of life in humans. The disadvantage is that as yet we do not have good techniques for measuring emotional states in animals. To have any indication of what an animal feels we have to take indirect measures, mainly by examining their behaviour.

Assessing welfare

Whereas the two lines of scientific approach to welfare assessment may at first appear rather different, they differ mostly in where the emphasis is placed – the immediate subjective experiences of the animal or its long-term biological functioning. When it comes to the actual methods used in the scientific assessment of animal welfare, there is actually good consensus between researchers. In the next section we will go briefly through the health, production and physiology measures and in this book on ethology, spend more time on the behavioural indicators that have been used to assess welfare.

Health and production as indicators of animal welfare

It is obvious that animal health is an important aspect of animal welfare. An animal with a broken leg, bleeding wound or other physical injury clearly has poorer welfare than an animal without that damage. Nevertheless, not all health issues are so clear and it is important to remember that the border between health and disease is often indistinct. Complete health, that is to say 100% disease free, probably doesn't even exist since the body is continually reacting to bacteria and viruses. Furthermore, even an unhealthy animal does not necessarily experience pain or distress. A dog with a small tumour probably does not experience any pain or distress and a chicken with weak bones will not experience pain until there is a fracture. None the less, the dog will probably suffer reduced welfare later if the tumour is not treated successfully, and the hen is at risk of reduced welfare since bones are often broken during handling and transport.

One simple conclusion, therefore, is that an unhealthy animal either has its welfare already reduced or is at risk some time in the future of having its welfare reduced. But, as will be explained later, there is more to good welfare than being healthy.

Just as historically a healthy animal has been regarded as an animal with good welfare, so it has been argued that a farm animal exhibiting high production has good welfare. But this view is criticized because production traits, e.g. number of offspring, growth rate, etc., have been the focus of intense selection for many generations. They differ between breeds and strains even under identical housing and management conditions and they can be easily manipulated by changes in diet, lighting schedules and so on. In fact, very high production itself can lead to problems, such as udder inflammation in high-producing dairy cows or leg problems in fast-growing broiler meat chickens and fattening pigs. These have been called production diseases and are a relatively new category of welfare problem resulting from high production. One final, albeit extreme, example to demonstrate the risk of using traditional production parameters as welfare indicators is the case of number of offspring. Using this measure, the animal with highest welfare would be a breeding bull at a semen station whose sperm are used to inseminate cows around the world. In fact, this bull may even be dead!

In summary, while poor production performance may be associated with poor welfare, good production *per se* is no guarantee of good welfare.

Physiological indicators of welfare

The most commonly used physiological indicators are those associated with the stress response outlined earlier in this chapter. Commonly used variables therefore include heart rate, corticosteroid levels and adrenal gland weight. Often, these measurements are made remotely using radiotelemetry or by measuring corticosteroids in non-invasive ways, from the saliva, urine or faeces, instead of from blood samples. Not disturbing the animal while taking the measurement is particularly important when the physiological response is rapid, as is the case with heart rate.

The frequently used interpretation is, the greater the stress the poorer the welfare. Usually this is true, but not always. The physiological response to short-term (acute) stress is different from that of long-term (chronic) stress because the system adapts and down-regulates. Also, some so-called stress responses can reflect positive experiences such as excitement and arousal.

To help overcome the problem of there being no obvious cut-off point where the level of stress can be said to be 'bad', some researchers have proposed that there is a welfare problem when the stress response is such that it leads to a change in the biological functioning of the animal and the animal enters a prepathological state. In short, it is argued that instead of measuring the stress response *per se*, the consequences of the stress are measured. Examples of such consequences can be immunosuppression or reduced reproductive success.

Behavioural indicators of welfare

The advantage of behavioural indicators of welfare is that they are the easiest to obtain and they probably reflect an animal's first attempts to cope with a less than optimal situation. Thus, responses such as huddling together when the temperature is too low or vocalizing when hungry probably occur long before there is a threat to welfare from hypothermia or starvation. Natural selection has led to mechanisms that operate earlier and earlier, thus giving us more sensitive indicators that welfare is at risk than the obvious indicators, such as injury and disease (Dawkins, 1998).

Comparison with normal behaviour

Early work on behaviour related to welfare was often associated with developing new housing systems and it was common practice to compare the time budgets of animals in the new system with those in the traditional system (a time budget refers to a quantification of how much time an animal allocates to different activities over the day). Comparisons were also made between animals kept under traditional systems with those kept under extensive conditions in outdoor enclosures. These extensive conditions were assumed to be better for welfare, and the standard upon which other systems could be compared; but there are several problems with such comparisons. That the time budgets of animals differ according to the environment in which they are kept is not surprising. Much more difficult is to interpret this difference in terms of the animal's welfare. If the range of behaviour patterns seen in one system is less than in the other, it may mean that the behaviour is not released by that particular environment, or it may mean that the animal is prevented from performing the behaviour.

An example of the first type may be anti-predator behaviour. Under extensive conditions poultry can give alarm vocalizations and escape to cover several times per day. If this behaviour is not seen under more confined conditions, we would not recommend that the caretaker takes steps to elicit it. The behaviour is triggered by external stimuli in the environment and there are unlikely to be welfare problems associated with its absence.

An example of the second type may be roosting behaviour in chickens, which is also in part an anti-predator behaviour, since predation losses are lower for birds roosting up off the ground. Given perches, birds will usually roost on them during darkness, otherwise they roost on the floor. Roosting behaviour is triggered by external stimuli, in this case the decreasing light intensity. A daily darkness phase is provided even in commercial poultry housing but there are no, or at least should not be any, real predators. The question is whether roosting up off the ground is still important for the birds and, if so, whether providing perches under commercial conditions would improve welfare. We could not have answered this question without further studies that showed that, indeed, birds are motivated to gain access to a perch for night-time roosting and so motivated for this behaviour. Thus, comparisons of behaviour in different environments can often highlight issues to be investigated experimentally.

One additional benefit of studies comparing the behaviour of the same animals in both extensive and intensive conditions, or even of modern breeds and hybrids with their ancestors, is the important information we obtain on where, when and how different behaviours are performed. Such studies help us understand some of the behaviours we see in captivity but have not been able to interpret. For example, it is common knowledge that many birds build nests. It is therefore likely that the restless behaviour a laying hen performs in the 1–2h before she lays her egg may be related to nest building. However, who would propose a similar explanation for the movement patterns shown by a sow about to farrow, if there had not been observations on nest building behaviour in nature? After all, no other ungulates are known to build nests.

Experimental studies of behaviour as welfare indicators

Experimental studies on behaviour have contributed greatly to our understanding of animal welfare. Rather than observing the undisturbed animal, as described above, an artificial situation is created where the behaviour of the animal helps us answer questions about its state. Imagine that we did not know how a cow with a painful leg would behave. To determine this, we could inject uric acid into a joint (to create a form of gout that we know is painful) and, in follow-up behaviour observations, we would observe that the animal limped, putting less weight on the treated limb. In future, when we saw an animal limping in the same way, we might conclude that it too was experiencing pain.

Now we do not need to perform this experiment, we already know the answer, but we can artificially create other states where we have less knowledge. Frustration is likely to result when an animal is prevented from performing a behaviour that it is motivated to perform. By experimentally creating a frustrating situation it is possible to study what behaviour patterns are shown by animals. Examples of frustration-related behaviour in chickens, for example, range from displacement preening when the level of frustration is mild, to increased aggression and stereotyped pacing in more frustrating situations. It is then possible to observe chickens in a variety of different situations to identify times of day or situations when there seems to be evidence of frustration, so that these situations can be studied in more detail. This idea of artificially creating emotional states has mostly been used to study frustration so far, but could probably be used to help investigate other states in animals. Recently, for example, similar methods have been used to show that animals develop expectations, for both positive and negative events. Signalling a later positive or negative event will evoke a range of various anticipatory behaviours, which can then be used in other situations to determine whether the animal finds some particular procedure rewarding or punishing.

Another methodology that has been used frequently is that of preference testing. Here, the animal is given a choice between two or more resources or situations and the assumption is that it chooses in its best interest. The concept of preference testing is simple, but it is important that the tests are well controlled if the results are to be reliable. For example, many of our domesticated animals are wary of unfamiliar environments and food. Thus, in a preference test the animal should have had equal experience of both choices. Animals also almost always choose to maximize short-term welfare over long-term – for example, a sweet tasty food over a nutritionally balanced one. Age, strain, previous experience and time of day will also influence the outcome.

Even with a well-designed preference test, though, there is rarely a 100% choice for one option over the other. Partial preferences may reflect that all animals rank the choices relatively similarly or that a proportion of the animals prefers one option whereas the other proportion prefers the other option. Partial preferences may even reflect that one choice is preferred on one occasion and the other the next time, depending on the state of the animal. One modification is to leave the animal in the test apparatus for a longer time and then compare the total time spent in each of the available compartments. While this may overcome some difficulties with the simple choice test, one must remember that time spent using a resource does not necessarily reflect the whole picture of its importance to the animal.

Despite the improvements in preference test experimental design, one cannot compensate for the fact that in preference tests the choice is always relative. That is to say, while one option may be strongly preferred over the other, both may be poor for welfare or both may be good. Training the animal to perform an operant response such as pecking a key or pushing a lever to gain access to a resource or an environment allows us to quantify the importance of a choice. The assumptions are that if an animal will work for the resource then it must be of at least some importance to it and, presumably, the harder the animal will work – i.e. more pecks or pushes – the more important it is (see Fig. 7.5).

A further refinement of this idea of quantifying how hard the animal will work for the resource is to incorporate it into consumer demand theory, used by economists to assess the importance of a commodity to human consumers. Commodities where demand decreases as price increases are said to have an elastic demand and those where the demand changes little as price increases are said to have inelastic demand. In humans, an example of the first may be fish and the second petrol. If the price of fish goes up people eat something else, but the majority of people still buy petrol for their car even if they complain about the price.

In animal studies, food is normally used as the standard example of an inelastic commodity, and a regression line of data on amount of work required to gain access to food with increasing price is usually rather horizontal. For other resources the regression line usually slopes downwards, showing that animals become less willing to obtain the resources as the amount of necessary work increases. By comparing the slopes of the lines one can rank the importance of different resources to animals (see Fig. 7.6). Just as this technique can be used to measure strength of motivation to obtain a resource, it can also be used to measure motivation to avoid something unpleasant.

Fig. 7.5. A laying hen pushes against a weighted door to open it and thus gain access to whatever resource is provided on the other side. The weight of the door, and so the effort needed to open it, can be manipulated systematically.

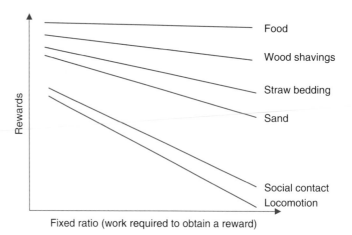

Fig. 7.6. Demand curves for various commodities in pigs. The pigs were required to perform an increased amount of work (numbers of snout presses on a bar) for access to a given commodity. The slope of the line indicates the relative importance of the commodity to the pig, a more horizontal slope indicating a higher importance (modified after Ladewig and Matthews, 1996).

7.5 Concluding Remarks

This chapter has shown that scientific studies of animal behaviour can provide some of the necessary keys for understanding the welfare of animals in captivity. Behaviour is often the first reaction used by an animal to adapt to a specific environment. Whereas data on physiological reactions and health are, of course, informative, they are often difficult to interpret in welfare terms without the necessary knowledge of accompanying behavioural reactions. For example, an elevated corticosteroid level may simply reflect an increased energy demand on the body, because the animal is actively playing, but in another situation it may signal that the animal is stressed, with an associated increased risk of disease and suffering. Observing behaviour can also help inform us about whether the animal is in pain or otherwise in a poor state of health.

Of course, the ultimate decision of whether a certain state of welfare is acceptable or not is an ethical one. It requires perspectives and arguments that cannot be obtained from scientific studies alone. However, any ethical discussion about what we ought or ought not do to animals should preferably be based on objective scientific knowledge about the state of the welfare of the animals.

References

Appleby, M.C. and Hughes, B.O. (1997) *Animal Welfare*. CAB International, Wallingford, UK, 316 pp.

Broom, D.M. (1996) Animal welfare defined in terms of attempts to cope with the environment. *Acta Agriculturae Scandinavica Section A Animal Science, Supplement* 27, 22–28.

Broom, D.M and Fraser, A.F. (eds) (2007) *Domestic Animal Behaviour and Welfare*, 4th edn. CAB International, Wallingford, UK.

Broom, D.M. and Johnson, K.G. (1993) *Stress and Animal Welfare*. Chapman & Hall, London, 211 pp.

Dawkins, M.S. (1998). Evolution and animal welfare. *The Quarterly Review of Biology* 73, 305–327.

Duncan, I.J.H. (1996) Animal welfare defined in terms of feelings. *Acta Agriculturae Scandinavica Section A Animal Science, Supplement* 27, 28–36.

Jensen, P. and Toates, F.M. (1997) Stress as a state of motivational systems. *Applied Animal Behaviour Science* 54, 235–243.

Ladewig, J. and Matthews, L.R. (1996) The role of operant conditioning in animal welfare research. *Acta Agriculturae Scandinavica Section A Animal Science, Supplement* 27, 64–68.

Mason, G.J. (1991) Stereotypies: a critical review. *Animal Behaviour* 41, 1015–1037.

Ruis, M.A.W., te Brake, J.H.A., van de Burgwal, J.A., de Jong, I., Blokhuis, H.J. and Koolhaas, J.M. (2000) Personalities in female domesticated pigs: behavioural and physiological indications. *Applied Animal Behaviour Science* 66, 31–47.

Sluyter, F., van Oortsmerssen, G.A., de Ruiter, A.J.H. and Koolhaas, J.M. (1996) Aggression in wild house mice: current state of affairs. *Behavioural Genetics* 26, 489–496.

Toates, F. (1995) *Stress – Conceptual and Biological Aspects*. John Wiley & Sons Ltd., Chichester, UK.

Weiss, J.M. (1971) Effects of coping behaviour in different warning signal conditions on stress pathology in rats. *Journal of Comparative and Physiological Psychology* 77, 1–13.

8   Human–Animal Relations

S. WAIBLINGER

8.1 The Human in the Animals' World

Throughout the history of the human species, animals have played an important part in human life, and vice versa. In the early hunter-gatherer/nomadic cultures, animals were viewed as prey, but also as dangerous predators. The same holds true for the animals' perspective: some species might have considered the human a predator, some a prey. For other species, humans might just have been a neutral part of the environment. When domestication began, the human–animal relationship developed towards a symbiosis in which the human provided protection from predators and food in exchange for animal products (food and fur) and power. For some individuals, e.g. animals 'adopted' by humans at an early age, the human was a social partner with whom they played and exchanged affiliative behaviour. Hediger (1965) first described the five possible roles (predator, prey, neutral part of the environment, symbiont, social partner) of humans for animals in captivity. Estep and Hetts (1992) suggested that some of these roles may not be mutually exclusive and that an animal probably perceives a human in terms of a combination of the above roles and according to the prevailing situational factors. An example would be a dog that may perceive a human (child) as prey (Miklósi, 2007) only in a particular situation.

Like their human counterparts, animals now take on a wide variety of roles: domestic animals being production animals (for food, wool, fur, leather); working animals (for instance, oxen or horses for transportation or ploughing the field; dogs for hunting, herding, protection; cats for catching rodents); animals used in sports (often just another form of working), in shows or as beloved pets. Obviously, differing relationships (between humans and animals) could be expected according to these roles. However, the relationship can also vary widely within those categories of animal utilization. The aim of this chapter is to discuss the reasons for variation in the relationship as well as its relevance, mainly regarding farm and pet animals. But, first, we shall deal with the definition of the human–animal relationship.

8.2 What is the Human–Animal Relationship?

When visiting different farms, e.g. working as a veterinarian, large differences in the behaviour of both animals and humans are striking: on some farms, animals are confident in the presence of humans and they can be examined and treated easily whereas, on others, animals are fearful of humans, treatment is difficult and one has to be careful to avoid being injured. Accordingly, on the former, the farmer talks to and strokes his/her animals and handles them calmly whereas, on the latter, farmers are barely aware of such gentle behaviour but are often impatient, shout at or hit their animals. Such variation has been shown in different species (Fig. 8.1). Similar differences can be found in dog–human or cat–human interactions.

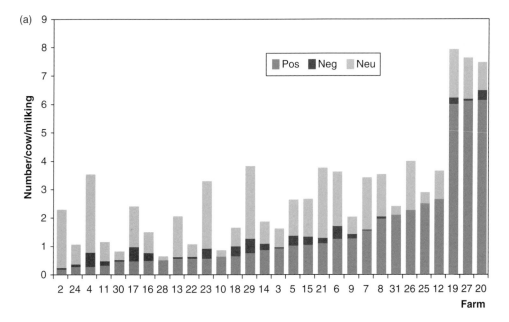

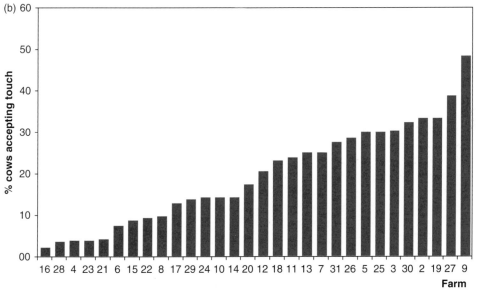

Fig. 8.1. Variation in human and animal behaviour on dairy farms. For exact definition of stockpeople behaviours and testing procedures see Waiblinger *et al.* (2006). (a) Number of different behaviours used by stockpeople per cow during one evening milking on 30 farms. Pos, friendly, positive interactions (stroking, touching, talking calmly); Neu, neutral or moderately negative interactions (slight hit with hand or stick, talking dominantly); Neg, clearly negative behaviour (strong hits with hand or stick, shouting, talking impatiently). (b) Percentage of cows that can be approached until being touched by an unfamiliar test person.

Obviously, the relationships of humans and animals differ distinctly in those examples. They are reflected in the behaviour of both animals and humans (see Fig 8.2). The basis for these behavioural differences is varying emotions and motivations during human–animal encounters due to the previous experience with each other. Thus, the history of previous interactions forms the foundation of the current relationship. The quality of the relationship then affects the nature and perception of future interactions. That means that a relationship is always dynamic: both positive and negative changes can happen throughout a lifetime, even though in some phases of life those changes are more likely to occur or to be longer lasting (as will be discussed below in more detail).

In sum, the human–animal relationship can be defined as the degree of relatedness or distance between the animal and the human, i.e. the mutual perception, which develops and expresses itself in their mutual behaviour (Estep and Hetts, 1992; Waiblinger *et al.*, 2006). The relationship can range on a continuum from poor or negative, where the human is perceived as frightening, and unpleasant emotions are involved during interactions, to good or positive, where the human is perceived as a social partner, and interacting with him/her is often pleasurable. The relative strength of pleasant or unpleasant emotions in the perception of humans constitutes an animal's relationship to humans and vice versa (Waiblinger *et al.*, 2006; Fig. 8.3). In this chapter we concentrate on the animal's perspective.

Individual or general relationships

In principle, a relationship develops between two individuals knowing each other – for instance, the caregiver and an animal in their care, or a pet animal and its owner.

Fig. 8.2. A mutual relationship is reflected in the behaviour of these cows (enjoying stroking or searching for contact) and the handler (stroking) (image courtesy of C. Menke).

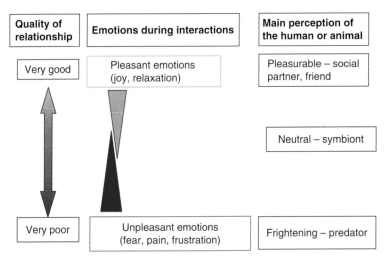

Quality of relationship	Emotions during interactions	Main perception of the human or animal
Very good	Pleasant emotions (joy, relaxation)	Pleasurable – social partner, friend
		Neutral – symbiont
Very poor	Unpleasant emotions (fear, pain, frustration)	Frightening – predator

Fig. 8.3. The two dimensions (unpleasant, pleasant) contributing to the human–animal relationship, and the corresponding perception of the human or animal. Increased levels of pleasant emotions improve the relationship and are characteristic for a very good relationship, and vice versa (modified from Waiblinger *et al.*, 2006).

Such relationships require mutual individual recognition. This is why, whereas in pet animals such personalized relationships are most common, in farm animals relationships are limited to systems enabling sufficient contact. Thus, large groups of farm animals – for example, a group of several thousand laying hens – impede the recognition of individual animals by the stockperson and thus individual relationships, even though specific individual animals may still receive special attention. However, animals often generalize their experience with familiar humans to unfamiliar humans (Bernstein, 2005; Waiblinger *et al.*, 2006; Fig. 8.4).

Farm animals that withdraw from their own stockperson at a great distance react in the same manner to an unknown human, whereas animals accepting being touched by their familiar stockperson also allow a strange person to do so. Humans as well may show generalized attitudes and behaviour towards animals of the same species or even in general (Hemsworth and Coleman, 1998). As long as individual recognition is not precluded, both individual relationships and generalization exist in parallel and influence animal behaviour. For example, lambs that had been bottle fed and had received gentle handling showed less isolation distress when a known or unknown shepherd was present, even though the effect was greater with a familiar shepherd. Handled piglets also interacted more with familiar and unfamiliar humans than non-handled ones, but made contact sooner and more often with the familiar handler and were less agitated when caught by her or him than by an unknown person.

Recognition of humans by animals

Individual relationships need individual recognition. The examples above have already shown that animals are able to differentiate between humans. But what cues do they use? When people are asked this question with respect to farm animals, they

Fig. 8.4. Gentle tactile contact in a generalized relationship: the lambs and the human had just met for the first time.

usually mention smell first. However, as for the recognition of conspecifics, mainly visual cues seem to be used whenever possible (Rushen *et al.*, 1999, 2001). Pigs, for example, can differentiate humans by the mere use of either olfactory, acoustic or visual cues, but they are worst with olfactory cues alone and best when they can use the three senses in concert. Like pigs, cattle are able to differentiate humans by their faces, but if clearer cues such as colour of clothes are available they make use of them. Previous experience with humans dressed in a specific colour during specific situations (be it pleasant or unpleasant) may then be translated to respective expectations when the same or other humans are dressed similarly. Sometimes, the sight of the clothes of a negative handler alone elicits stress responses (Waiblinger *et al.*, 2006). Sheep can not only differentiate the faces of different humans, but also remember them 2 years later (Kendrick, 2006).This agrees with anecdotal reports of bulls remembering an aversive person more than 1 year later.

8.3 Factors Affecting the Quality of the Animal's Relationship to Humans

The previous history of human–animal interactions, i.e. the quality and quantity of the interactions and the period of life when they happened, contains the main determinants of an animal's relationship to humans. However, the genetic background of the animal, as well as the physical and social environment for rearing and interacting with humans, is also influential. These factors are discussed in this section.

In domestic animals, it is mainly the human who determines the possibilities of interactions, their type and timing within a given environment. Often he/she also provides the environmental conditions for interactions. Therefore, the human contributes

S. Waiblinger

to a great extent to the animal's developing relationship with him/her. The main factors underlying human behaviour are personality and attitudes, but empathy, knowledge, experience and the actual situation (e.g. time pressure, workload) are also influential. These human factors have been reviewed, for instance, by Serpell and Paul, (1994), Hemsworth and Coleman (1998), Podberscek *et al.* (2000) and Spoolder and Waiblinger (2008), and will not be considered here.

Quality of human–animal interactions

Human–animal interactions can involve visual, tactile, olfactory and auditory perception. An animal may perceive an interaction as negative, neutral or positive. As mentioned above, this perception is influenced by the existing relationship with humans. The simple approach of a human can be perceived by different animals in any of the three ways mentioned above, according to previous experience. However, some interactions are aversive by their very nature because they are painful or otherwise distressing (e.g. dehorning, beak trimming, forceful hitting by hand or stick). Shouting and hitting are perceived as aversive, as shown in cattle by aversion learning techniques and preference tests or handling studies, where acute and chronic stress responses were found (Waiblinger *et al.*, 2006). Frequent or predominant use of negative interactions such as hitting (even with low force) or shouting, inducing pain or discomfort, increases animals' fear of humans, whereas neutral interactions (such as visual presence) allow habituation to humans with decreased levels of fear. Positive interactions (such as stroking, feeding, gentle talking) reduce fear and enhance confidence in humans reflected in behaviour (more approach and less avoidance) and physiology (lower heart rate and cortisol levels) (Boivin *et al.*, 2003; Hemsworth, 2003; Waiblinger *et al.*, 2006).

Different sensory channels provide different possibilities. Even though visual presence alone can be powerful in habituating animals to humans and thus in improving the relationship, (gentle) tactile interactions are more powerful, especially when provided in a manner imitating intraspecific social grooming (McMillan, 1999; Schmied *et al.*, 2008a). Gentle stroking elicits quite distinct behavioural responses (e.g. lowering of eyelids, hanging of ears), indicating a state of relaxation. In several species it was also shown to trigger oxytocin release and thereby to lead not only to immediate effects of a lower heart rate, lower blood pressure and lower cortisol levels, but also to long-term positive effects on social bonding and health (Uvnäs-Moberg, 1997; Uvnäs-Moberg and Petersson, 2005). However, a certain level of confidence in humans seems necessary to elicit such positive physiological responses.

Gentle contact may not be perceived as positive by every individual. Besides the existing relationship, the prevailing motivation and environment are likely to influence the perception. It would be interesting to know whether forced stroking could also have positive effects. Recent results in cattle and horses suggest this to be the case, at least in animals used to some form of human contact (Ligout *et al.*, 2008; Schmied *et al.*, 2008b). In newborn foals, forced handling had the opposite effect (Hausberger *et al.*, 2008). However, the reaction of the handled animal may be a reliable indicator of its perception: the early-handled foals showed strong escape attempts, whereas defensive behaviours were rare in the example of Schmied's force-stroked cattle. Furthermore, subtle differences in the quality of interaction (pressure during stroking accompanied by a gentle voice or not) could be important for these effects (Hennessy *et al.*, 1998).

Food is often recommended as establishing relationships with animals. Research suggests that, even though offering food is actually a good start to a relationship, it has a less rewarding component than other positive interactions such as gentle tactile interactions or playing, and is thus not sufficient for maintaining a (positive) relationship. Cats, for instance, seem to know which family member is most likely to provide food and direct their interactions towards that person when they are hungry. Nevertheless, they are likely to be just as affectionate to other family members at other times (Bradshaw, 1992). Another experiment showed that, although feeding first attracts cats to the person feeding them, the preference for this person over a non-feeder (outside the feeding situation) ceased after a few days – other interactions such as playing, petting, etc. are required to cement a newly founded relationship (Karsh and Turner, 1988).

Gentle, soothing talking is often used by farmers as well as in experiments combined with other forms of contact, but there are few investigations on the effects of talking alone. In cats a larger effect was found when handling included talking (Bernstein, 2005; see also below in Section 8.8). For reducing stress, interactions should further be predictable and controllable. For instance, the random incorporation of negative interactions within a positive contact had similar, negative effects on welfare to a consistent regime of negative handling (Boivin *et al.*, 2003).

Quantity of human–animal interactions

Apparently, the amount of interaction is also crucial. Fear responses to humans can be caused not only by a learnt negative association due to aversive interactions, but also by an absence of habituation to human contact. With an increasing amount of neutral or positive contact the animal's relationship to humans will improve. For example, calves handled daily during the first 3 months of life and running with their mother from the 4th month were easier to handle, showed less flight behaviour, accepted longer tactile contact and, importantly, were never aggressive in a 'docility test' at the age of 8 months compared with calves running with their mother from birth onward. Kittens handled for over 1 h/day went directly to a familiar person, climbed into his or her lap, purred and either played or slept. In contrast, a kitten that had been handled less, e.g. a mere 15 min/day, tended only to approach and to head rub and then move off (Karsh and Turner, 1988).

8.4 Sensitive Periods

In dogs and cats, there is clear evidence that the sensitive period of socialization is extremely important for their later relationship with humans; in other domestic animals too, sensitive periods seem to be important. Socialization refers to the process by which an animal develops appropriate social behaviour towards conspecifics (Karsh and Turner, 1988). During the socialization period, the animal forms primary social relationships or social attachments; in the wild, it attaches itself to conspecifics;

S. Waiblinger

in domesticated animals, contact during this period with humans and other non-conspecifics allows the formation of relationships to them (Serpell and Jagoe, 1995). Being raised throughout the socialization period with non-conspecifics only (e.g. cross-fostering of kittens and pups with the opposite species only) leads to a preference of this species for later positive social interactions (Serpell and Jagoe, 1995). Moreover, hand-reared ruminants and pigs display courtship behaviour preferably or exclusively towards humans, depending on the duration of isolation from conspecifics (Sambraus and Sambraus, 1975). Likewise, if raised with their conspecifics, the amount and type of contact with humans (or other non-conspecifics) during the socialization period will more or less determine the later social behaviour towards humans (or other non-conspecifics) in dogs and cats.

Adequate socialization of kittens or puppies to humans is a precondition for a positive pet–owner relationship and is beneficial for preventing the development of inappropriate behaviour later in life. It is difficult to compensate a lack of contact with humans during the sensitive period of socialization by later interactions. In cats, for example, animals handled in the sensitive period of 2–7 weeks of age are generally friendly to people, whereas non-handled kittens will be extremely difficult to socialize at a later date (Karsh and Turner, 1988). In dogs, the primary socialization period is thought to run from about the 3rd to the 12th week after birth, with a peak of sensitivity between 6 and 8 weeks (Serpell and Jagoe, 1995). However, in dogs, regular social reinforcement until the age of around 8 months is necessary to keep socialization intact, as dogs well socialized at 3 months of age will become fearful again without it. Socialization of dogs later in life is also very difficult and requires considerable patience (Serpell and Jagoe, 1995).

The amount of contact, as well as the number of different humans during the sensitive period, is likely to affect the later relationship with humans. As mentioned above, kittens handled for more than 1 h/day show more positive behaviour and stay longer in contact with a familiar person compared with less-handled kittens. Cats handled by one person only can be held longer by the familiar person than by a stranger, but cats with experience of four handlers will stay with any person for the same amount of time. Also, kittens with one handler only were more affectionate to humans, particularly to their handler, purred more often and played for longer times. Kittens handled by five people were more outgoing and showed least fear of strangers (Karsh and Turner, 1988; Bradshaw, 1992). Interestingly, the type of contact or the sensory channels involved also seem important: when handling included talking, the socialization of kittens (handling during the early development) was especially effective (Bernstein, 2005).

In farm animals too, the period of an animal's life during which human contact occurs can be important, even though the early socialization period seems less crucial. In herbivores, the times soon after both birth and weaning seem to constitute sensitive periods, and are especially effective in improving the relationship, i.e. the same amount and type of contact has stronger impact than if given later (Rushen *et al.*, 2001; Boivin *et al.*, 2003; Waiblinger *et al.*, 2006). For example, heifers handled gently in the first 2 weeks after birth or weaning accepted being stroked and fed by the hand of a human to a much greater degree than when handled in the same manner 6 weeks after birth or weaning. The latter, however, were still less fearful than animals not handled at all. What these periods have in common is that they are related to the formation of, or changes in, social relationships (to the mother, the

Fig. 8.5. Early positive contact such as play is crucial for a good relationship later in life.

offspring or peers). Again, also in farm animals, contact limited to those sensitive periods helps to improve the animal–human relationship, but a prolonged period of regular handling seems to be necessary to maintain it (Boivin *et al.*, 2003; Hausberger *et al.*, 2008). Furthermore, in foals, it was even argued that there is no clear evidence for the existence of sensitive periods throughout their development to facilitate the establishment of a foal–human bond. Two-year-old non-handled animals become as familiar as handled animals (Hausberger *et al.*, 2008). However, they were exposed daily to caretakers bringing food, which may be sufficient for primary socialization. Thus, the relative importance of handling during sensitive periods and prolonged or lifelong regular handling in farm animals is still ambiguous.

In sum, for both pet and farm animals the best relationship can develop and be maintained when (positive) contact with humans takes place during sensitive periods early in life and regular, positive interactions also happen later on (see Fig. 8.5). This is in agreement with intraspecific social relationships, where the strongest bonds are formed earlier in life and are regularly reinforced by positive social interactions. It has to be stressed, however, that an adequate socialization with conspecifics is important for later life, and separation of the pup or kitten from its mother and siblings too early can also cause problems later.

8.5 Genetic Background and Personality

A major characteristic of domestication is a change in animals' reactions to humans, which can be summarized by decreased fear and greater 'tameness'. The powerful role

S. Waiblinger

of genetics can be seen in recent selection experiments in silver foxes, where behaviour towards humans changed quickly within one generation due to selection for tameness. In later generations, behaviour changed more and more to patterns resembling those of dogs (e.g. wagging tail, licking the human's hand) (Price, 2002; Miklósi, 2007). In domesticated animals, the genetic influence on human–animal interactions, and thus the relationship, becomes evident from differences between breeds and individual differences within breed (Serpell and Jagoe, 1995; Turner, 2000; Rushen *et al.*, 2001; Price, 2002; Boissy *et al.*, 2005; Hausberger *et al.*, 2008).

Results from different species suggest that the influence of the genetic background on the animals' relationship with humans is due to underlying changes in different personality traits, mainly fearfulness/boldness, aggressiveness, curiosity and sociability (Turner, 2000; Bernstein, 2005; Boissy *et al.*, 2005; Miklósi, 2007). Personality can be defined as an individual's unique set of traits, which is relatively stable over time and affects how it interacts with the environment. In cats, for example, paternity was shown to influence animals' behaviour towards humans via influences on the trait of boldness. Reported heritability of fear of humans or reactivity to handling varies widely in cattle, sheep and pigs, from low heritability ($h^2 \approx 0.1$) to quite high heritability (0.53). Presumably, such variations result from differences in test procedures measuring reactions to humans. The fact that the estimated heritability of traits related to reactions to handling or to humans can change with the age of the animals (lower heritability in adult versus young animals) points to the importance of experience.

Genetic background and experience with humans interact in the development of the animal–human relationship. Cats being socialized to humans in the first 3 months of life and having a father friendly to humans were more likely to show relaxed behaviour and less likely to show defensive behaviour towards a stranger (Bernstein, 2005). In some cases, personality seems to be quite strong: in cats, some individuals seem resistant to changes of their original type of interaction with humans, i.e. some kittens are quite friendly to humans from the beginning and remain so, whereas others remain quite fearful despite handling (Bernstein, 2005). Individual personality types amongst cats having a comparable socialization history were one of the most significant factors influencing their behaviour towards people in experimental settings (Turner, 2000).

8.6 Social Learning: Social Environment

In social species such as our domestic animals, social influences on interactions and the relationship of a given individual with humans are not surprising. In groups of animals, fear of humans may be transmitted via behavioural and/or olfactory cues. Likewise, relaxed and bold behaviour of some individuals towards humans may also facilitate approach for others. The mother's reactions to humans may especially be transmitted to its young through social facilitation, as shown in goats, quails and horses (Hausberger *et al.*, 2008). Young goats' responses towards humans were shown to correlate with the positive responses of an adult goat, especially the dam, towards humans. In horses, gentle handling of the mother (during the first 5 days after birth) revealed a long-lasting (at least 1 year) improvement in the foal's relationship with the familiar human as well as a generalization to unfamiliar humans

(Hausberger *et al.*, 2008). Thus, tame mothers, but probably also other social models, may play a crucial role in the development of a positive human–animal relationship in their young.

These effects, however, have received more attention only recently, whereas earlier studies mostly concentrated on the effects of direct handling of the young (as discussed above). There, research in sheep and cattle showed that the presence of the dam could limit the efficiency of early human contact given to lambs or calves on later reaction to their stockperson (Boivin *et al.*, 2003). In cats, the presence of the (calm) mother facilitates interactions of kittens with humans after a short period when kittens prefer to stay close to the mother (Karsh and Turner, 1988). In horses, a direct comparison of early handling of foals in the presence of the mother, exposure to a motionless human and gentle handling of the mother (all during the first 5 days after birth) revealed a long-lasting improvement in the foal's relationship with humans, only for those foals having observed their dam being handled. Exposure to a motionless human has some limited, short-lasting improving effect (Hausberger *et al.*, 2008). In cats, the mother may even have another influential role when preventing kittens from human contact during sensitive periods, by hiding them.

8.7 The Human as a Social Partner

The social nature of a human–animal relationship, where the human is a social partner of the animals, is apparent from animals behaving equally to both humans and conspecifics. This comprises essentially affiliative behaviours, but also dominance-related behaviours and species-specific communication, and the ability to provide social support (Rushen *et al.*, 2001). In 1950 Scott (cited in Karsh and Turner, 1988) stated that 'the behaviour patterns exhibited by dogs toward human beings are essentially the same as those exhibited toward dogs'. Turner (2000) states that the human–cat relationship is indeed a 'two-way partnership, with both parties adjusting their behaviour to that of their partners'. Cats and dogs seek the proximity of humans, direct affiliative social behaviour towards them, initiate grooming (stroking) interactions or play – obviously enjoying these interactions.

This is also true for farm animals with good relationships (see Figs 8.2 and 8.4), but there is far less scientific evidence. In the 1970s, the Fulani herdsmen and their cattle were described as forming a dual-species social system fulfilling the following characteristics: following early mutual socialization of cattle and humans, the herdsmen are integrated into the herd by taking the role of a leader, a dominant (humans clearly exhibit dominance over the animals and learn how to do so early in childhood) and a companion (humans invest much time in stroking animals in body regions where the latter often lick each other) (Lott and Hart, 1979). As a consequence, animals can be called by their herdsmen for restraint (Hinrichsen, 1979).

Furthermore, the human can give animals social support and reduce stress in aversive situations, e.g. isolation from conspecifics or a veterinary procedure (McMillan, 1999; Waiblinger *et al.*, 2006). While attachment is not necessarily involved in those behaviours and effects, dogs and cats are often attached to their owner and this may occur also in farm animals that have had intensive contact with humans, especially hand-reared animals (Karsh and Turner, 1988; Boivin *et al.*, 2003; Waiblinger *et al.*, 2006). Recently, it has been argued that the behaviours of dogs and human children are parallel in their pattern of attachment. Both dogs and children develop similar behavioural malformations, which might relate to the attachment relationship (e.g. separation anxiety) (Miklósi, 2007).

8.8 Human–Animal Communication

Herding a group of cows, moving pigs, riding a horse, calling a cat to come for feeding, a dog guiding a blind human or a dog and his/her owner playing together – all these situations involve human–animal communication. Different sensory channels are involved to a different degree according to the situation. Vocalizations, body postures, facial expressions, movements and the release of pheromones all have communicative value. When living in a social group, effective communication is extremely important, and thus animals have developed special communicative behaviours (e.g. threatening postures or facial expressions, submission…), which are all inborn or learned. Communication between species bears the enhanced risk of misinterpretation of signals. However, different species can learn to understand each other, at best via mutual socialization (see Fig. 8.5). In human–animal interactions, imitating species-specific animal signals has been recommended for effective control of farm animals as well as in the training of dogs. Miklósi (2007) suggests that, besides the use of species-specific signals, the consideration of the rules of intraspecific social interactions also seems to improve human–animal communication.

However, there seem to be species-independent features of signalling behaviour. For instance, threatening or submissive behaviours are often associated with making the body appear taller or smaller, respectively, and with gazing directly at the opponent or avoiding gaze (Eibl-Eibesfeldt, 1999). Indeed, avoiding direct gaze and only glancing at the animals, as well as a smaller body shape such as a quadrupedal posture, squatting, sitting or lying, allow a closer and quicker approach as compared with standing upright and gazing directly towards the animal (Kendrick, 1998; Waiblinger et al., 2006; Spoolder and Waiblinger, 2008). Many humans use such signals unconsciously (e.g. squatting to make animals approach). Moreover, the direction of approach, movement style (quick or slow, deliberate), and even more subtle cues such as muscle tension and breathing patterns, may all play a role (Boivin et al., 2003; Waiblinger et al., 2006). Dogs react appropriately to a stranger approaching in a friendly or threatening way, seemingly reacting to the whole pattern of human behaviour (Miklósi, 2007).

In addition, with respect to the characteristics of vocalization, species-independent communalities seem to exist. There is evidence that some physical properties of sound have consistent, species-independent effects on the response of an animal receiver and that humans use these features for effective communication with dogs (McConnell, 1991). Short, rapidly repeated broadband notes are used to stimulate motor activity, whereas longer, continuous narrowband notes are used to inhibit motor activity by animal trainers with a wide geographical and linguistic background. Dog pups learn better to come when trained with short, rapidly repeated sounds. Interestingly, the barking of dogs carries emotional information for the human receiver – humans can categorize barks from different situations over chance level independently of having owned a dog or not (Miklósi, 2007).

Wemelsfelder showed in her research on 'qualitative behavioural assessment' that humans can assess quite reliably aspects of animals' emotional state by observing them in their environment and by using both the whole range of animal behaviour, without quantifying it, and the universal features of body postures, movement, tension and facial expressions (Wemelsfelder, 2007; Napolitano *et al.*, 2008).

Besides such universal communicative signals, human–animal communication can reach a very sophisticated level of 'understanding', depending on the level of socialization and intensity of human–animal contact as well as on individual and breed differences (for a review see Miklósi and Soproni, 2006; Miklósi, 2007). One famous example was the horse 'Kluger Hans', which was able to 'calculate' mathematical tasks. By sensing and reacting to changes in humans' body tension, he stamped the correct number of times with his foreleg. This is possibly also a good example of humans unconsciously emitting signals.

Dogs were shown to be able to comprehend human communicative signals and be aware of human attentional states. For example, dogs follow the pointing or gaze of humans to find hidden food or toys. They use the direction of attention of humans to obtain food from the appropriate person and to perform forbidden actions, but they can also communicate the location of a hidden target to the human. A particular Border collie was found to be able to associate 340 spoken words with objects or specific persons. Again, genetic potential and experience interact: working dogs are better than breeds not selected for work, and trained working dogs are more skilled than pet working dogs. Interestingly, cooperative hunting breeds perform better than non-cooperative breeds.

The cooperative aspect of human–animal relationship and communication becomes apparent in guide dogs and their owners. During work, guide dogs have the ability to exchange roles interactively with the human as the initiator of actions. Dogs also adapt their communicative behaviour to the visual state of their owner: guide dogs used sonorous mouth-licking behaviour when interacting with their blind owners when compared with pet dogs having sighted owners during a test situation where food was unreachable for the dogs (Serpell and Paul, 1994; Gaunet, 2008).

Not only dogs understand the communicative signals of humans; cats and goats were shown to do so, even though the latter seemed to perform less well. Interestingly, wolves with the same history of socialization with humans during the first 3–4 months of life are mostly unable to comprehend and use pointing gestures at the same age as do dogs, but they are able to later in life. This supports the hypothesis that dogs (and other domesticated species) have been selected for enhanced sociocognitive abilities for living in human social settings, but also points to the effects of socialization and experience.

8.9 Practical Relevance: the Significance of Human–Animal Relationships for Welfare

In both farm and pet animals, research has shown that the actual human–animal relationship has a huge impact on animal welfare and performance, as well as on human safety and welfare. As per definition, in a poor relationship interactions with (negatively interacting) humans elicit negative emotions such as fear, causing physiological stress reactions (acute and chronic) and behaviours with negative consequences for risk of injuries, accidents and ease of handling (due to flight or defensive behaviour, aggression), for health (via depression of the immune system or lesions, e.g. claw lesions), reproduction, performance and meat quality. A good relationship not only minimizes such stress reactions, but even has an anti-stress potential, alleviating other sources of stress such as a necessary veterinary treatment or restraint (Seabrook and Bartle, 1992; Hemsworth and Coleman, 1998; McMillan, 1999; Rushen *et al.*, 1999; Boivin *et al.*, 2003; Hemsworth, 2003, 2004; Waiblinger *et al.*, 2006; Hausberger *et al.*, 2008).

In dogs and cats a good relationship – and thus adequate socialization as well as adequate interactive behaviour of the owner – helps to prevent behavioural problems, including aggression; both the human and the animal benefit from this. Behavioural problems are often the reason for sheltering animals or even euthanasia. In both dogs and horses, the use of punishment for training can cause stress and aggression towards humans, while dogs whose owners use positive reinforcement (in a correct way) show better compliance.

A special aspect is relevant, especially for students of animal behaviour: the potential influence of animals' relationship to humans on outcomes of experimental studies caused by observer or caregiver influences (Davis and Balfour, 1992).

In conclusion, the existing knowledge of the possibilities for a good quality of relationship for animals with humans should be further implemented to improve animal and human welfare. However, many questions are still unanswered: for example, with reference to the most effective ways of communication, the perception and importance of special interactive behaviours, and the role and potential of social learning for improved relationships in farm animals.

References

Bernstein, P.L. (2005) The human–cat relationship. In: Rochlitz, I. (ed.) *The Welfare of Cats*. Springer, Dordrecht, the Netherlands, pp. 47–90.

Boissy, A., Fisher, A.D., Bouix, J., Hinch, G.N. and Le Neindre, P. (2005) Genetics of fear in ruminant livestock. *Livestock Production Science* 93, 23–32.

Boivin, X., Lensink, J., Tallet, C. and Veissier, I. (2003) Stockmanship and farm animal welfare. *Animal Welfare* 12, 479–492.

Bradshaw, J.W.S. (1992) *The Behaviour of the Domestic Cat*. CAB International, Wallingford, UK, 219 pp.

Davis, H. and Balfour, A.D. (1992) *The Inevitable Bond – Examining Scientist–Animal Interactions*. Cambridge University Press, Cambridge, UK.

Eibl-Eibesfeldt, I. (1999) Grundriss der vergleichenden Verhaltensforschung. *Piper, München – Zürich* 8, 932.

Estep, D.Q. and Hetts, S. (1992) Interactions, relationships, and bonds: the conceptual basis for scientist–animal relations. In: Davis, H. and Balfour, A.D. (eds) *The Inevitable Bond – Examining Scientist–Animal Interactions*. Cambridge University Press, Cambridge, UK, pp. 6–26.

Gaunet, F. (2008) How do guide dogs of blind owners and pet dogs of sighted owners (*Canis familiaris*) ask their owners for food? *Animal Cognition* 11(3), 475–483.

Hausberger, M., Roche, H., Henry, S. and Visser, E.K. (2008) A review of the human–horse relationship. *Applied Animal Behaviour Science* 109, 1–24.

Hediger, H. (1965) *Mensch und Tier im Zoo*. Albert-Müller Verlag, Rüschlikon, Zurich, Switzerland.

Hemsworth, P.H. (2003) Human–animal interactions in livestock production. *Applied Animal Behaviour Science* 81, 185–198.

Hemsworth, P.H. (2004) Human–animal interactions. In: Perry, G.C. (ed.) *Welfare of the Laying Hen.* CAB International, Wallingford, UK, pp. 329–343.

Hemsworth, P.H. and Coleman, G.J. (1998) *Human–Livestock Interactions: the Stockperson and the Productivity of Intensively Farmed Animals.* CAB International, Wallingford, UK.

Hennessy, M.B., Williams, M.T., Miller, D.D., Douglas, C.W. and Voith, V.L. (1998) Influence of male and female petters on plasma cortisol and behaviour: can human interaction reduce the stress of dogs in a public animal shelter? *Applied Animal Behaviour Science* 61, 63–77.

Hinrichsen, J.K. (1979) Mensch-Tier-Beziehung bei afrikanischen Rindernomaden. *KTBL-Schrift* 254, 103–110.

Karsh, E.B. and Turner, D.C. (1988) The human–cat relationship. In: Turner, D.C. and Bateson, P. (eds) *The Domestic Cat: the Biology of its Behaviour.* Cambridge University Press, Cambridge, UK, pp. 159–177.

Kendrick, K.M. (1998) Intelligent perception. *Applied Animal Behaviour Science* 57, 213–231.

Kendrick, K.M. (2006) Brain asymmetries for face recognition and emotion control in sheep. *Cortex* 42, 96–98.

Ligout, S., Bouissou, M.F. and Boivin, X. (2008) Comparison of the effects of two different handling methods on the subsequent behaviour of Anglo-Arabian foals toward humans and handling. *Applied Animal Behaviour Science* 113, 175–188.

Lott, D.F. and Hart, B.L. (1979) Applied ethology in a nomadic cattle culture. *Applied Animal Ethology* 5, 309–319.

McConnell, P.B. (1991) Lessons from animal trainers: the effect of acoustic structure on an animal's response. In: Bateson, P. and Klopfer, P. (eds) *Perspectives in Ethology. Human Understanding and Animal Awareness.* Plenum Press, New York, London, pp. 165–187.

McMillan, F.D. (1999) Effects of human contact on animal health and well-being. *Journal of the American Veterinary Medical Association* 215, 1592–1598.

Miklósi, Á. (2007) *Dog Behaviour, Evolution, and Cognition.* Oxford University Press, Oxford, UK, 274 pp.

Miklósi, Á. and Soproni, K. (2006) A comparative analysis of animals' understanding of the human pointing gesture. *Animal Cognition* 9, 81–93.

Napolitano, F., De Rosa, G., Braghieri, A., Grasso, F., Bordi, A. and Wemelsfelder, F. (2008) The qualitative assessment of responsiveness to environmental challenge in horses and ponies. *Applied Animal Behaviour Science* 109, 342–354.

Podberscek, A.L., Paul, E.S. and Serpell, J.A. (2000) *Companion Animals and Us: Exploring the Relationships Between People and Pets.* Cambridge University Press, Cambridge, UK.

Price, E.O. (2002) *Animal Domestication and Behavior.* CAB International, Wallingford, UK.

Rushen, J., Taylor, A.A. and de Passillé, A.M. (1999) Domestic animals' fear of humans and its effect on their welfare. *Applied Animal Behaviour Science* 65, 285–303.

Rushen, J., de Passillé, A.M., Munksgaard, L. and Tanida, H. (2001) People as social actors in the world of farm animals. In: Keeling, L.J. and Gonyou, H.W. (eds) *Social Behaviour of Farm Animals.* CAB International, Wallingford, UK, pp. 353–372.

Sambraus, H.H. and Sambraus, D. (1975) Prägung von Nutztieren auf Menschen. *Zeitschrift für Tierpsychologie* 38, 1–17.

Schmied, C., Boivin, X. and Waiblinger, S. (2008a) Stroking different body regions of dairy cows: effects on avoidance and approach behavior towards a human. *Journal of Dairy Science* 91, 596–605.

Schmied, C., Waiblinger, S., Scharl, T., Leisch, F. and Boivin, X. (2008b) Stroking of different body regions by a human: effects on behaviour and heart rate of dairy cows. *Applied Animal Behaviour Science* 109, 25–38.

Seabrook, M.F. and Bartle, N.C. (1992) Human factors. In: Phillips, C. and Piggins, D. (eds) *Farm Animals and the Environment.* CAB International, Wallingford, UK, pp. 111–125.

Serpell, J. and Jagoe, J.A. (1995) Early experience and the development of behaviour. In: Serpell, J. (ed.) *The Domestic Dog. Its Evolution, Behaviour and Interactions with People.* Cambridge University Press, Cambridge, UK, pp. 79–102.

Serpell, J. and Paul, E. (1994) Pets and the development of positive attitudes to animals. In: Manning, A. and Serpell, J. (eds) *Animals and Human Society: Changing Perspectives.* Routledge, New York, pp. 127–144.

Spoolder, H.A.M. and Waiblinger, S. (2008) Pigs and humans. In: Marchant-Forde, J.N. (ed.) *The Welfare of Pigs.* Springer, New York, pp. 211–236.

Turner, D.C. (2000) Human–cat interactions: relationships with, and breed differences between, on-pedigree, Persian and Siamese cats. In: Podberscek, A.L., Paul, E.S. and Serpell, J.A. (eds) *Companion Animals and Us: Exploring the Relationships Between People and Pets.* Cambridge University Press, Cambridge, UK, pp. 257–271.

Uvnäs-Moberg, K. (1997) Physiological and endocrine effects of social contact. In: Carter, S., Lederhendler, I.I. and Kirkpatrick, B. (eds) *The Integrative Neurobiology of Affiliation.* New York Academy of Science, pp. 146–163.

Uvnäs-Moberg, K. and Petersson, M. (2005) Oxytocin, ein Vermittler von Antistress, Wohlbefinden, sozialer Interaktion, Wachstum und Heilung. [Oxytocin, a mediator of anti-stress, well-being, social interaction, growth and healing]. *Zeitschrift für Psychosomatische Medizin Psychotherapie* 51, 57–80.

Waiblinger, S., Boivin, X., Pedersen, V., Tosi, M., Janczak, A.M., Visser, E.K. and Jones, R.B. (2006) Assessing the human–animal relationship in farmed species: a critical review. *Applied Animal Behaviour Science*

Wemelsfelder, F. (2007) How animals communicate quality of life: the qualitative assessment of behaviour. *Animal Welfare* 16, 25–31.

PART II: SPECIES-SPECIFIC BEHAVIOUR OF SOME IMPORTANT DOMESTIC ANIMALS

Editor's Introduction

For the second part of the book, we turn our attention to the actual behaviour seen in some of the more common and economically important domestic animals. With the basis of the theoretical approach in the first part, it is now possible to go into detail about what animals actually do.

The species treated in the book have been selected because they are the most common domestic animals, and also those about which most research has been performed. For most of the species we will examine, there is considerable knowledge of the behaviour of the ancestors, and of the ways in which these animals react and behave under different husbandry conditions.

For each species, the authors have attempted to provide information on a number of different aspects. The origin and domestication history is given some emphasis in all chapters. Sections on social behaviour and communication, foraging and feeding habits, mating behaviour and other aspects of reproduction are then treated, although the exact order and content may vary somewhat between chapters, dependent on the relevance for the species treated. Behavioural ontogeny is a further subject which is treated as far as information is available.

Most of the text is devoted to the normal behaviour of the species but, in several chapters, we also examine a selection of applied problems. These cover typical issues raised by housing and husbandry conditions, including common behavioural disorders. Where information is available, the authors have attempted to provide some ideas about possible prevention or treatment of the types of problems described.

With the information provided in this second part of the book, it is hoped that students of animal behaviour will be more than well equipped to differentiate between normal and abnormal behaviour when animals are observed under practical husbandry conditions.

9 Behaviour of Fowl and Other Domesticated Birds

J.A. Mench

9.1 Origins

Many avian species have been domesticated for use by humans (Appleby *et al.*, 2004), including ground-dwelling fowl (chickens, turkeys, guinea fowl, quail and pheasants), waterfowl (ducks and geese), the flightless ratites (ostrich, emu and rhea) and pigeons. Chickens (*Gallus gallus domesticus*), ducks (and especially Pekin ducks, *Anas platyrhynchos*) and turkeys (*Melagris gallopavo*) are the most commonly used species worldwide, with an estimated 56 billion chickens, 2.7 billion ducks and 650 million turkeys raised commercially for meat or eggs annually.

Chickens were derived from the red junglefowl, *Gallus gallus*, modern forms of which are found in India, Burma (Myanmar), Malaysia, Thailand and Cambodia. Junglefowl were domesticated over 8000 years ago in South-east Asia, and were then taken by humans north into China, and from there to Europe. Domestic fowl were established in many European countries by 100 BCE (before the Common, i.e. Christian, Era, or BC), and were brought to the Americas around 1500 CE (Common Era, or AD). During the initial stages of domestication the fowl was probably valued mainly as a sacrificial or religious bird, or for cockfighting. However, the Romans created specialized breeds, including highly productive egg layers, and formed a complex poultry industry. This industry collapsed with the fall of the Roman Empire, and did not resume on a large scale until the 19th century, when there was an emphasis on breeding fowl for both ornamental and production traits.

Modern breeds are derived mainly from the two types of fowl developed during this period, Asiatic and Mediterranean. These were then hybridized in the 20th century for commercial use to create egg-laying and meat strains, referred to as laying hens and broiler chickens, respectively. Laying strains have been intensely selected for traits such as early onset of egg laying, a high rate of egg production, egg quality characteristics (shell strength, egg size) and food conversion efficiency. Broiler chickens have similarly been selected for food conversion efficiency but, in addition, intense selection pressure has been applied for rapid growth, a high ratio of white to dark meat (larger breast size) and meat yield, resulting in a chicken that reaches full adult body size in only 6 weeks, as compared with 17 weeks for an egg-laying chicken.

Turkeys are native to the Americas, and wild populations are still widespread throughout North and Central America. They were domesticated about 2000 years ago in Mexico. The early Spanish explorers brought turkeys back to Spain, and they were quickly distributed throughout Europe. The domesticated form of the turkey was then taken to North America by the European settlers in the 17th century. Breeding originally focused on plumage characteristics for show purposes. By the 20th century, however, selection programmes had begun to focus on meat production and, as a consequence, turkeys have been selected for many of the same traits as broiler chickens. In addition, there was selection against the dark plumage of the wild turkey in order to remove the hormone melanin from the feather follicles, which improved the appearance of the carcass.

The Pekin duck was derived from the mallard, a species widely distributed throughout North America and Eurasia. Mallards were probably domesticated in Asia about 4000 years ago, and were farmed by the Romans for meat. The mallard has given rise to a large number of domesticated breeds of ducks, including breeds that have a high rate of egg laying. However, the most common breed is the Pekin from China, which is used for meat and feather (down) production. Like broilers and turkeys, commercial Pekin ducks have been selected for rapid growth and good feed conversion.

Other, less common, avian species domesticated for food production include the Muscovy duck (*Cairina moschata*), geese (*Anser anser*), guinea fowl (*Numidia meleagris*), pigeons (*Columbia livia*) and pheasants (*Phasianus colchicus*). The Muscovy is a forest duck native to South and Central America; its date of domestication is uncertain, although it was certainly domesticated by the 15th century. Muscovy are used for meat, and are also hybridized with Pekin to produce moulard ducks, which are used for the production of fatty livers (*foie gras*). Geese have probably been domesticated several times, but most domestic forms are derived from the greylag goose, found in China and South-east Asia. They are produced for meat, feathers and *foie gras*. Guinea fowl were domesticated about 5000 years ago in Africa and are produced commercially there for meat and eggs. Pheasants are native to Central Asia, and were probably distributed across Europe as semi-domesticated birds by the Romans. They have only recently begun to be kept in close confinement and bred for meat production.

The most recently domesticated species are the Japanese and bobwhite quail (*Colinus* sp.), from Japan and North America, respectively, and the ratite species (ostrich, emu and rhea) from Africa and Australia. The ostrich (*Struthio camelus*) is the most widely used of the ratites. These are all multi-purpose species, used for both meat and eggs and sometimes for feathers. All have undergone domestication during the last hundred years.

Around the world, many of these species are raised on a limited scale for local consumption. Under these conditions they would often be kept outdoors in small flocks, sometimes with indoor shelter provided for inclement weather. For large-scale commercial production, however, all except the largest or less domesticated species (like ratites and pheasants) are typically housed indoors to provide better environmental control, to allow automation of feeding, watering, egg collection and manure disposal, and for the purpose of disease prevention.

There are two basic types of commercial housing: floor systems and cages (see Fig. 9.1). Cages are the primary type of commercial housing used worldwide for mature egg-laying hens and quail. Meat-type birds (broilers, turkeys, ducks) and breeding flocks are typically housed in floor systems, with the birds being housed on flooring made of wire, slats, litter (e.g. wood shavings) or some combination of these. Because of concerns about the restriction of natural behaviour in some developed countries, however, an increasing number of caged egg-laying hens are being housed in so-called furnished cages, which contain perches, nest boxes and dust baths, in floor systems or even outdoors in free-range systems. A more detailed review of the different kinds of housing systems in commercial use for poultry can be found in Appleby *et al.* (2004).

Fig. 9.1. The conventional (battery) cage is the most common system worldwide for housing hens (a), although newer designs of cages (furnished cages) containing nest boxes, perches and dust baths, are becoming increasingly common in some European countries, as are free-range (b) production systems. Meat-type birds like ducks (c), turkeys and broiler chickens are usually housed in floor systems containing thousands or tens of thousands of birds; some laying hens are also kept in such systems (image of ducks courtesy of Mike Turk).

This chapter will focus mainly on the behaviour of chickens, turkeys and ducks, not only because they are the most commonly used species, but because their behaviour has been best studied. However, as discussed in Chapter 2, many domesticated species share behavioural traits that rendered them suitable for domestication, so many of the topics covered in this chapter will apply also to other domesticated birds.

9.2 Social Behaviour

All poultry species are highly social, although their wild ancestors show different forms of social organization (Mench and Keeling, 2001). Some live in small relatively stable groups. Junglefowl, for example, typically live in groups comprising one male and several females, with other males being solitary or living in small, all-male groups. Turkeys may also live in mixed-sex groups during the breeding season, but the rest of the year most commonly stay in all-male or all-female groups. Ostriches also live in stable groups comprised of family members of different ages and sexes, except during the breeding season when they form breeding pairs or trios with two females and one male. The social groups of quail, ducks and geese are less stable in that group composition changes after the breeding season, or after migration to breeding and feeding grounds.

As in all social species, there is extensive communication among group members. Poultry species have excellent colour vision and acute hearing, and communication occurs via visual and vocal signals. Postures and displays are used to signal threats and social submission, and particularly elaborate displays are given during courtship (see below). Features of the head and neck are particularly important in some species for social recognition and communication. In chickens, comb size and colour are affected by sex hormone levels and are indicators of social status. Turkeys have a pale featherless neck area that changes colour to red and blue during social interactions, and the fleshy, pendant snood of male turkeys becomes engorged and enlarged during aggression and courtship.

Most poultry species also have an extensive vocal repertoire – the exceptions are ostriches and Muscovy ducks, which rarely vocalize. Calls can serve a variety of functions, including warning about approaching predators, decreasing the distance between flock-mates (contact calls), signalling threat or submission, or attracting offspring or other flock members to food. The most striking vocalizations are the ones that males make during territorial defence – the 'crows' of roosters and Japanese quail, the 'booms' of ostriches and the 'whistle' of bobwhites. These calls carry long distances and thus are an effective way for a male to defend a territory without having directly to confront the males on neighbouring territories.

Calls can also be used to signal social dominance. The crow of roosters is individually distinctive, and roosters use the quality and rate of crowing to assess the dominance status of other roosters. Social dominance hierarchies (peck orders) are a common feature of social organization in poultry. As the name implies, these orders are formed and maintained by aggressive pecking directed towards the head region of more subordinate birds, which, in turn, show submissive behaviour (see Fig. 9.2). In mixed-sex flocks, males and females generally develop separate dominance hierarchies and rarely show aggression towards one another. In established flocks, aggressive and submissive behaviours are usually subtle and difficult to observe, although sometimes there can be chasing, pecking and even fighting, particularly among males.

Despite periodic aggressive interactions, even when poultry are kept in large areas where they could avoid one another they tend to cluster together. The tendency to form groups rather than to live independently evolved primarily for protection against predators. Even in the absence of predators, however, poultry often move together as a flock. They also tend to synchronize their behaviours, either because of circadian effects (see below) or due to social facilitation, with birds 'copying' one another's behaviour. Observing and copying other birds may allow a bird to exploit new resources (e.g. new food types or sources) or learn new behaviours – for example, chickens that have an opportunity to observe a trained hen peck a key to obtain

Fig. 9.2. Social hierarchies in poultry are maintained through aggressive and submissive interactions. The bird on the left is adopting a confident and aggressive posture, while the bird on the right is showing submission.

J.A. Mench

food learn this task much more quickly than chickens that do not have this experience (Johnson *et al.*, 1986).

In commercial conditions poultry may be kept in extremely large flocks containing thousands or even tens of thousands of birds (see Fig. 9.1), or in small groups but under crowded conditions where the birds cannot easily avoid one another (like cages). Except for breeding flocks, birds are also typically kept commercially in single-age, single-sex flocks instead of in the mixed-age and -sex flocks that would be the norm in the wild. As is true of many domesticated animals, poultry are relatively tolerant of these variations in social groupings. However, social behaviour problems can arise in commercial settings (Mench and Keeling, 2001). Turkeys, for example, can be particularly aggressive towards one another in commercial flocks, sometimes pecking each other so severely that the head wounds result in death (Sherwin and Kelland, 1998). They are therefore usually kept in dim lighting to reduce pecking behaviour.

Social problems are sometimes seen also in female flocks. Subordinate laying hens may be pecked continuously by other birds – they have heads and combs scarred from pecking, poor body condition and spend much of their time trying to avoid other birds. This problem is particularly noticeable in moderate-sized groups, since in small groups the hens know one another and have a stable dominance hierarchy, whereas in large groups they seem to develop non-aggressive social strategies for establishing dominance. Even in the small groups typical of cages, however, more dominant hens may sometimes prevent subordinate hens from accessing the feeder to such an extent that the subordinate hens lose body condition and stop producing eggs (Cunningham and van Tienhoven, 1983). In non-cage systems, housing roosters with the hens can help to decrease problems with bullying of subordinates (Odén *et al.*, 2000), since roosters suppress aggression amongst hens.

The most problematical social behaviours seen in commercial flocks are feather-pecking and cannibalism (Appleby *et al.*, 2004). These are abnormal behaviours, and are more common in large than small flocks. Feather-pecking is the pecking and pulling of feathers from another hen. Unlike aggressive pecks, which are directed towards the head, feather pecks are directed towards body regions like the areas near the vent and preen gland, and the wings, back and tail. The pecking movements involved resemble feeding pecks rather than aggressive pecks, and there is evidence that feather-pecking is redirected foraging behaviour and can be reduced by providing foraging materials. For the recipient, having feathers pulled out is painful, and birds with large areas of exposed skin have more difficulty regulating their body temperature.

Cannibalism, which involves the pecking and tearing of the skin and underlying tissues of another bird, is an even more serious problem and one that can result in extremely high mortality in flocks. Cannibalism sometimes follows from feather-pecking, but can also arise independently. One situation in which it starts is when a hen has just laid an egg and her cloaca is still partly everted. Other hens are attracted to peck at this area, especially if the skin becomes broken and bleeds, leading to further pecking and consumption of the flesh.

Despite a great deal of research, the exact causes of feather-pecking and cannibalism are still not completely understood. The extent of these problems can vary enormously from one flock to another even when the management of the flocks is very similar, and flocks do not necessarily experience both problems at the same time. Factors affecting the incidence of both types of behaviours include stocking density, genetics, lighting, whether or not foraging materials are provided, and feed composition and form. One of the most important factors for fowl is group size, and these problems are more common in non-cage systems where large numbers of hens are housed together.

Methods used to decrease problems with feather-pecking and cannibalism include: (i) keeping the birds in dim or red light (which makes it more difficult for them to see wounds); (ii) fitting them with devices (like contact lenses or 'spectacles') that reduce their ability to see; or (iii) beak or bill trimming. This last is the most commonly used method. It involves removing about one-third to one-half of the upper beak or bill, and is effective in reducing the damage that the birds can inflict upon one another. This practice has been criticized, however, because it causes short-term (and sometimes long-term) pain (Hester and Shea-Moore, 2003) and, as a consequence, it has been banned in some countries. A promising alternative is genetic selection to reduce the propensity for birds to show feather-pecking and cannibalistic behaviour. This approach has been quite successful in reducing these problems in caged laying hens (Muir, 1996), but it is presently unclear how well it will work for non-caged flocks, where these problems are most severe.

9.3 Foraging, Feeding and Drinking

Under natural conditions, poultry and their wild ancestors have a varied diet. All species feed on plants and, depending upon the species, may consume grasses, shrubs, roots, leaves and berries. They may also eat invertebrates, and some species even eat small vertebrates like lizards and mice.

Foraging for food is a very important part of the natural behaviour of poultry – for example, under semi-natural conditions fowl devote a large proportion of their day to foraging activities. Components of foraging behaviour include ground pecking, ground scratching and grazing. Ducks also filter edible items out of the water by dabbling, using their bills.

Under commercial conditions, concentrated feed is provided in troughs and is usually readily available throughout the day. However, poultry still spend considerable time foraging if an appropriate foraging substrate, like loose litter, is provided. This indicates that they are motivated to forage even in the absence of a need to do so.

In housing environments without a foraging substrate, such as conventional cages, birds do not have access to such material and may instead spend a substantial portion of time manipulating the food in the feed trough by flicking it back and forth using movements of the beak. This can lead to the food being tossed out of the trough and wasted, which is an economic issue for egg producers. Methods used for reducing feed wastage include beak trimming (discussed above), which removes the hook from the beak and thus prevents food from being caught under the hook and flicked, and altering the feeder to prevent the birds from performing foraging movements – whether this results in frustration, however, is unknown.

There has been a considerable amount of research on food selection by chickens (Appleby et al., 2004). Chickens have a well-developed sense of taste, and reject potential food items that are acidic, bitter or extremely salty. Visual and tactile cues are also important – both chicks and adults preferentially peck at items of particular colours and have a preference for pecking at and ingesting small, rounded particles.

J.A. Mench

These preferences are, however, affected by their experience with particular food types, and may also vary depending upon the birds' nutritional status. Under natural conditions, wild birds are faced with an array of food items that differ in nutritional quality, from which they must select a diet to meet their nutritional needs. Domestic birds show a similar ability to select a nutritionally adequate diet when given a choice of different feedstuffs, and will adjust their intake of protein, energy, minerals and vitamins (Hughes, 1984). Poultry do, however, develop a preference for the food to which they are accustomed, and become reluctant to consume new foods that differ in colour, taste or texture from their typical food (food neophobia). A major change in diet can thus cause them to reduce their feed intake, and consequently result in a reduction in growth or egg production.

Social factors have an important influence on feeding behaviour. Poultry show a propensity to feed as a group – that is, to feed synchronously – and feeding behaviour can be triggered by the sight or sound of other birds feeding. Individually caged hens will even synchronize their feeding behaviour with that of the hens in the neighbouring cages. If insufficient feeder space is provided for all birds to feed simultaneously, they may need to desynchronize their behaviour in order for all birds to consume enough feed.

Bouts of feeding are alternated with brief bouts of drinking. Chicks are attracted to pecking at shiny items, which in a natural setting would probably result in them finding, pecking at and ingesting water. Poultry drink by scooping up water in their beaks or bills, and then raising their heads so that the water runs down the oesophagus (see Fig. 9.3). In commercial settings poultry may be offered water in trough or bell-type drinkers or cups, which allows them to perform this natural drinking motion. However, to prevent water spillage, which can lead to the litter and manure becoming wet, it is now more common for birds to be given nipple drinkers instead of troughs or cups. Nipple drinkers require the birds to drink in an unnatural way,

Fig. 9.3. Birds normally drink by scooping up water and then elevating their heads slightly so that the water moves down their oesophagus by gravity (a). Because of concerns about water leaking and making the litter or manure wet, it is common to provide the birds with nipple waterers, which require them to learn to drink in an unnatural position (b) (images courtesy of Houldcroft *et al.*, 2008).

although they can learn to do so and in fact develop various strategies for activating the nipple and consuming water. Poultry do become accustomed to using particular drinker types and may have difficulty drinking if they are moved to a house where there is a novel type of drinker – in this case, they may have to be shown where the drinkers are located, or be taught to activate them.

9.4   Body Maintenance

It is important for birds to keep their feathers clean and in good condition. The feathers provide a covering and insulating layer that helps to maintain body temperature and prevent injury to the skin. Birds perform two primary behaviours to maintain plumage condition: preening and dust or water bathing. During a preening bout, the bird uses its beak and face to distribute oil from a gland at the base of the tail, the uropygial gland, through its feathers (see Fig. 9.4a). The beak is also used at this time to align the barbs of the feathers, and to dislodge external parasites like mites and ticks.

Water bathing by ducks, and dust bathing by other poultry species, serves to remove dirt and excess oil from the feathers and improves feather structure. Dust bathing has been studied more than water bathing. A dust bathing bout begins with the bird lying down and pulling loose substrate close to its body (see Fig. 9.4b). The bird then rubs itself on the substrate and shakes its wings and body to toss the material on to its back and work it through the feathers. The dust bath ends with the bird standing up and shaking its body to remove the excess loose material and to realign its feathers. Fine materials such as sand or peat are preferred for dust bathing, probably because they are superior at penetrating the feathers to reach the downy portions.

To fully perform dust or water bathing birds must be provided with either loose material or water that is deep enough for them to immerse themselves. However, waterfowl raised commercially are rarely given bathing water due to cleanliness and disease problems, and poultry housed in cages are generally not provided with loose material. This is a welfare concern and has led to a great deal of research on the motivational aspects of dust bathing behaviour. Laying hens housed in cages without litter material will carry out dust bathing movements on the wire floor of their cages,

Fig. 9.4. Ground-dwelling poultry maintain their feathers in good condition by preening to distribute oil through their feathers (a), and by dust bathing to work loose material through their feathers, which removes any excess oil (b) (image of preening courtesy of Cleide Falcone, and of dust bathing, Anna Lundberg).

J.A. Mench

so-called 'sham' dust bathing, although it is unclear whether this behaviour actually satisfies their motivation to dust bathe.

9.5 Diurnal Rhythm

Most behaviours in poultry do not occur at random, but instead show distinct daily rhythms. All poultry species are diurnal, and hence sleep at night and are active during the daylight hours. The behaviours that show the strongest rhythms are those related to feeding, egg laying, mating, grooming and sleeping. Feeding behaviour occurs in bouts, or meals, with most bouts occurring in the first few hours after the lights go on (or dawn, when there is natural lighting), and then again late in the day before the lights go off (or dusk, when there is natural lighting). Because poultry do not typically feed at night, the morning feeding peak allows them to refill their food storage organ (the crop), which has become depleted overnight. The crop is then filled again for the night during the feeding peak late in the day. In laying hens, this late afternoon feeding peak also correlates roughly with the start of eggshell calcification, at which time there is an increase in the hen's calcium requirements.

Egg laying in many poultry species occurs at about the time the lights go on. This is approximately 24 h after ovulation, and reflects the period of time required for the egg to form as it moves through the oviduct. This means that the pre-laying behaviours associated with selecting a nest site and building a nest also show a strong diurnal rhythm, since they occur shortly before the egg is laid. The daily cycle of mating in breeding birds is also related to the egg-laying cycle, because the hen's fertility is lower around the time of oviposition. Mating is therefore most frequent in fowl in the afternoon because their eggs are laid in the morning, but most frequent in quail in the morning and evening because they lay their eggs in the afternoon.

Like feeding, preening behaviour occurs primarily in the morning and late afternoon. However, dust bathing behaviour takes place early in the afternoon. On average, fowl dust bathe for about 30 min every 2–3 days (Vestergaard, 1982). The strongly diurnal rhythm of this behaviour can be used to commercial advantage – to prevent hens from laying their eggs in the dust bath, which makes the eggs difficult to collect, dust bathing material can be provided only in the afternoons, after the peak of egg laying.

Although periods of rest may occur throughout the day, most resting and sleeping occurs at night. During periods of deep sleep poultry tuck their heads under their wings, although they may also doze with their heads upright or on the ground and their eyes closed or partially open. The smaller poultry species are very vulnerable to ground and aerial predators when asleep, so they prefer to roost in areas that offer some protection, for example on the water (waterfowl), in dense cover (quail) or on elevated roosting areas (chickens and turkeys). Roosting is a highly motivated behaviour – hens will push through a weighted door to gain access to a perch at night (Olsson and Keeling, 2002), and 100% of hens in commercial houses will perch on elevated perches at night if they are given the opportunity to do so.

A number of factors can affect the diurnal rhythm of behaviour (Appleby *et al.*, 2004). For example, feeding rhythms are affected by the form and density of the diet, because these influence how long it takes the bird to consume an adequate meal. They are also affected by genetics, with birds selected for a high rate of feed intake and weight gain, like broilers and turkeys, more likely to eat throughout the day and even at night. Egg-laying can be delayed by factors such as human disturbance or social interference from other birds (for example, a dominant hen preventing a subordinate hen from entering a nest box) – in this case, laying may not be accompanied by pre-laying behaviour and the egg may just be laid while the hen is performing other activities during the day.

The most important factor affecting all diurnal rhythms, however, is the light cycle. Poultry have photoreceptors not only in their eyes but also in the pineal gland in their brain, and thus are very sensitive to light stimulation. Light acts as a 'time keeper' (a *zeitgeber*), controlling the circadian rhythms of behaviour. The commercial poultry industry raises birds under many different kinds of light cycles. Some of these provide only very brief periods of either light or darkness, or levels of illumination that are so low that there is no distinct light/dark cycle. Broiler chickens and turkeys, for example, are often kept in very dim lighting to decrease activity and hence promote more rapid growth. This can lead to a marked change in their diurnal rhythms of behaviour. Figure 9.5 shows the patterns of foraging and preening behaviour in broiler chickens given 16 h of light and 8 h of darkness per day, but kept in either dim (5 lux) or brighter (50 lux) lighting during the light period. The broilers kept in dim lighting show a much less distinct rhythm of behaviour, and tend to distribute their foraging and preening behaviour more evenly throughout the day and night.

9.6 Sexual Behaviour

The ancestors of poultry species show a variety of different types of mating systems, ranging from promiscuity (polyandry, polygyny or both) to monogamous pair bonds that last for one or more seasons (Mench and Keeling, 2001). Individual males may set up territories to which females are attracted during the breeding season, or instead may associate with a harem of females year-round. Alternatively, males and females may congregate during the breeding season at special breeding grounds, called leks, where the females select the males with which they will mate.

Fowl have a harem polygynous mating system, with a dominant male maintaining a territory and mating with the females that live in his territory. Turkeys may also form harems under some circumstances, but usually mate in leks. Related male turkeys occupy a breeding site on the lek and attract females to their group by displaying; the dominant male turkey at a particular site usually secures most of the matings during the height of the breeding season. Bobwhite quail form mating pairs during the spring and are usually monogamous during a particular breeding season. Ducks and Japanese quail are more variable in their behaviour; they sometimes form monogamous pair bonds, but also may mate promiscuously. Wild geese form long-term pair bonds in the wild, but the domesticated forms often mate promiscuously. The mating behaviour of ratites is particularly complex – they show both polygyny and polyandry during the breeding season, but also form short-term pair bonds.

Even in apparently promiscuous mating systems, birds (and particularly female birds) do not mate randomly but are selective about their choice of mates. Mate selection has been best studied in junglefowl. Female junglefowl use a variety of physical characteristics to assess the suitability of an unfamiliar male – including comb colour,

J.A. Mench

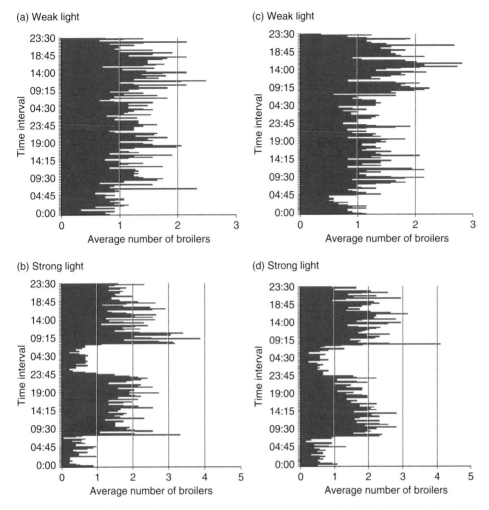

Fig. 9.5. Diurnal rhythms of behaviour are strongly influenced by the light cycle. Broiler chickens reared with a light cycle that has a strong light–dark contrast show a much more pronounced rhythm of foraging (b) and preening (d) behaviour than do broilers reared in dim lighting with little light–dark contrast (a, c), as shown by the average numbers of broilers performing those behaviours at different times of day over a 48 h period.

eye colour, spur length and comb size. Comb size is one of the most important cues and, since males that are less healthy have smaller combs, this may be one method females have for assessing the male's fitness (Zuk *et al.*, 1990). If the hen is familiar with the male, his dominance status and courtship behaviour are more important selection factors than his physical features.

Once potential mates are selected, courtship consists of a chain of stimulus–response patterns between the male and the female (see Fig. 9.6). The male usually is the obvious initiator of the sequence, but females do encourage courtship by approaching males or, in the case of female ostriches, engaging in displays. Male courtship displays are often very elaborate, involving noises, vocalizations, conspicuous

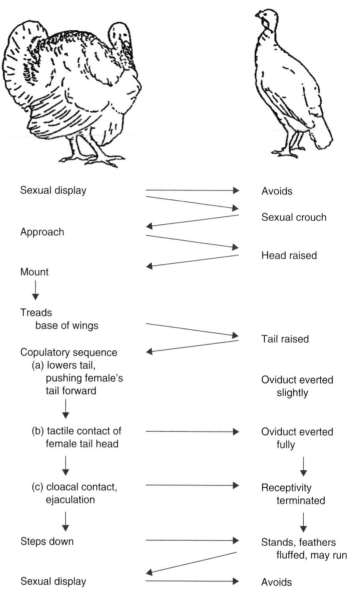

Fig. 9.6. Mating behaviour involves an often elaborate sequence of behaviours between the male and female. When males, like this tom (left), display by posturing, spreading their tail and showing neck colour changes, the female (right) may avoid the male or crouch to indicate that she is receptive. If she crouches, the turkeys then engage in a copulatory sequence that ends with the tom inseminating the hen (from Hale *et al.*, 1969).

postures and spreading the feathers or wings to make the body look larger. Males sometimes also show colour changes or enlargement of certain body parts, such as the snood of male turkeys. The female signals her receptiveness to mating by crouching and everting her cloaca, which allows the male to mount her and mate.

J.A. Mench

In commercial systems, mating patterns may change because of the ways in which the birds are managed (Appleby *et al.*, 2004). Poultry are typically allowed to mate naturally, but are often kept in large groups. Under these circumstances, factors such as social dominance are much less likely to affect mate selection, since the birds probably cannot easily recognize one another. The ratio of males to females in commercial settings is designed to encourage promiscuous mating to ensure the maximum fertilization of eggs. In such settings both males and females may mate a number of times during the day even though such frequent mating is not necessary to produce fertilized eggs, since hens can store viable sperm in a specialized gland for days to weeks. The exception to natural mating under commercial conditions is in turkeys, which have been selected for such large breast size that the males can no longer get close enough to the females to copulate. For this reason, turkey hens are artificially inseminated with semen collected from the toms in the flock.

9.7 Egg Laying, Incubation and Behaviour at Hatching

The behaviour associated with egg laying is expressed relatively inflexibly, probably because it is under genetic and physiological control. In fowl, pre-laying behaviour is triggered by the release of oestrogen and progesterone from the follicle after it is ovulated. Once it is triggered, the hen is very strongly motivated to find an appropriate nest site in which to lay her egg. A hen examines different potential sites, selects one and then commences to make a rudimentary nest by using her feet and rotating her body to create a hollow. During this time the egg has continued to develop as it moves down the oviduct, and it is laid after nesting behaviour is complete. The entire sequence, from nest searching to oviposition, takes about 1–2 h. The searching phase of the sequence may be extended when hens do not have access to a suitable nesting site, as in conventional cages.

Eggs laid outside the nest can be a significant problem in commercial production, since such eggs are difficult to collect and can become cracked and dirty, affecting their hatchability and economic value. So-called floor laying is variable both within and between systems, but there is increasing understanding of the factors that affect this problem (Appleby *et al.*, 2004). Nest box design is important to hens. They tend to prefer dark, secluded areas as they would in the wild, as these areas are better protected from predators, although their preference is affected by early experience. However, nest box design probably plays more of a role in determining which box a hen selects to lay her eggs in, rather than whether she lays them in a nest box or on the floor. What may be more important to the latter is how accessible the nest boxes are – most nest boxes in commercial houses are above ground level, and hens reared with no experience of perching may not be able to reach them easily and hence, instead, lay their eggs on the floor (Appleby *et al.*, 1986).

Under natural circumstances, once a clutch of eggs is laid the parents sit on the eggs nearly continuously to keep them warm until they hatch. In chickens and turkeys the female alone incubates the eggs, but in other species (pigeons, ostriches and bobwhite quail) both parents participate in incubation, while in some species (e.g. rheas) only the male incubates the eggs. Incubation behaviour in females is triggered by release of the hormone prolactin, which also causes the cessation of egg laying. However, genetically selecting commercial hens to produce a large number of eggs has also inadvertently resulted in selection against incubation behaviour. Therefore, most commercial strains of hens simply leave the nest after oviposition, and the pre-laying and laying cycle then begins again the next time an egg is to be laid.

Social interactions between parents and offspring occur in birds even before hatching (Rogers, 1995). Calls made by the developing embryos stimulate the incubating parent to turn the eggs or to return to the nest to resume incubation. Embryos also respond to the calls and behaviour of the parent with calls that further influence the parent's behaviour. Since the incubation of eggs is automated in commercial poultry production, these parent–offspring interactions are absent. However, the embryos also vocalize to one another, which causes the development of the less advanced embryos to accelerate and thus ensures that all of the eggs hatch at around the same time.

All domestic poultry except pigeons are precocial when they hatch, which means that their eyes are open, they are covered with down, they are mobile and they require little parental care. Under natural conditions, however, chicks and ducklings would stay close to one or both of their parents for several weeks after hatching. This closeness is maintained through a process called imprinting. Imprinting is a special form of learning that occurs during a sensitive period shortly after hatching, where the newly hatched bird instinctively follows the first moving objects it sees and thus learns the characteristics of its parent. In the absence of a parent or other adult bird, as is the case in commercial poultry production where typically only birds of the same age are housed together, chicks can imprint on other chicks, humans or even objects.

9.8 Care of Offspring

During the first few weeks after hatching chicks, ducklings and poults are not able to maintain their body temperature well. The hen (or in the case of ratites both the female and the male) therefore 'broods' them to keep them warm by covering them with her body and wings. Although precocial young otherwise require little parental care, the parents do play a role in protecting young chicks from danger and teaching them about various aspects of the environment. The mother hen helps chicks to learn how to go up to roost on branches at night. Domestic fowl chicks and turkey poults are attracted to edible foods (and steered away from inedible or toxic foods) by the pecking and vocalizations of the hen. In commercial flocks young birds are kept in single-age groups and thus there are no parental influences on their behaviour. While fowl chicks readily explore and find food and water themselves, or by copying the behaviour of other chicks that have already found food and water, turkey poults sometimes 'starve out', meaning that they fail to start eating, presumably because of the lack of maternal stimulation. Placing conspicuous and attractive stimuli near the feeder, such as flashing coloured lights, can help to attract the poults to the feeder (Lewis and Hurnik, 1979).

9.9 Offspring Development

Poultry become fully independent of their parents a few weeks or months after they hatch. During this time, their behaviours are developing into the forms seen in adult birds. As they explore and learn about foods, pecking at inedible items like sand decreases and pecking at edible items increases. They learn to recognize suitable dust bathing substrates, and the different elements that make up a dust bathing bout begin to appear in their behavioural repertoire, with full dust baths finally being performed when they are several weeks of age.

Social behaviour is also developing at this time. Young chicks frolic and spar with one another, play behaviour that resembles adult chasing and fighting. Chickens begin to peck others aggressively when they are as young as 2 weeks of age, although submissive behaviours are not shown until a few weeks later. Aggression is not apparent in turkeys until they are much older, about 3 months of age. In either case, these interactions lead to the formation of male and female dominance hierarchies in the flock, although these may not become stable until the birds reach sexual maturity.

In addition, developing birds learn the characteristics of appropriate mates for normal sexual activity through a process called sexual imprinting, which occurs during a sensitive period prior to sexual maturity. If males and females are reared separately, as sometimes occurs in commercial production, a lack of ability to undergo sexual imprinting can cause problems with mating later on in breeding flocks, since the birds may form strong homosexual pair bonds or show reduced mating behaviour (Appleby *et al.*, 2004). However, unlike some other bird species, in poultry it is not necessary for males to be present in order for the females to become sexually mature – the reason that hens can be kept in all-female groups for egg production.

References

Appleby, M.C., Maguire, S.N. and McRae, H.E. (1986) Nesting and floor laying by domestic hens in a commercial flock. *British Poultry Science* 27, 75–82.

Appleby, M.C., Mench, J.A. and Hughes, B.O. (2004) *Poultry Behaviour and Welfare*. CAB International, Wallingford, UK.

Cunningham, D.L. and van Tienhoven, A. (1983) Relationship between production factors and dominance in White Leghorn hens in a study of social rank and cage design. *Applied Animal Ethology* 11, 33–44.

Hale, E.B., Schleit, W.M. and Schein, M.W. (1969) The behaviour of turkeys. In: Hafez, E.S.E. (ed.) *The Behaviour of Domestic Animals*, 2nd edn. Williams and Wilkins, Baltimore, Maryland, pp. 22–24.

Hester, P.Y. and Shea-Moore, M. (2003) Beak trimming egg-laying strains of chickens. *World's Poultry Science Journal* 59, 458–474.

Houldcroft, E., Smith, C., Mrowicki, R., Headland, L., Grieveson, S., Jones, T.A. and Dawkins, M.S. (2008) Welfare implications of nipple drinkers for broiler chickens. *Animal Welfare* 17, 1–10.

Hughes, B.O. (1984) The principles underlying choice of feeding behaviour in fowls – with special reference to production aspects. *World's Poultry Science Journal* 39, 218–228.

Johnson, S.B., Hamm, R.J. and Leahey, T.H. (1986) Observational learning in *Gallus gallus domesticus* with and without a conspecific model. *Bulletin of the Psychonomic Society* 24, 237–239.

Lewis, N.J. and Hurnik, J.F. (1979) Stimulation of feeding in neonatal turkeys by flashing lights. *Applied Animal Ethology* 5, 161–171.

Mench, J.A. and Keeling, L.J. (2001) The social behaviour of domestic birds. In: Keeling, L.J. and Gonyou, H.W. (eds) *Social Behaviour in Farm Animals*. CAB International, Wallingford, UK, pp. 177–210.

Muir, W.M. (1996) Group selection for adaptation to multi-hen cages: selection program and direct responses. *Poultry Science* 15, 349–359.

Odén, K., Vestergaard, K.S. and Algers, B. (2000) Space use and agonistic behaviour in relation to sex composition in large flocks of laying hens. *Applied Animal Behaviour Science* 62, 219–231.

Olsson, I.A.S. and Keeling, L.J. (2002) The push-door for measuring motivation in hens: laying hens are motivated to perch at night. *Animal Welfare* 11, 11–19.

Rogers, L.J. (1995) *The Development of Brain and Behaviour in the Chicken*. CAB International, Wallingford, UK.

Sherwin, C.M. and Kelland, A. (1998) Time budgets, comfort behaviours, and injurious pecking of turkeys housed in pairs. *British Poultry Science* 39, 325–332.

Vestergaard, K. (1982) Dust-bathing in the domestic fowl – diurnal rhythm and dust deprivation. *Applied Animal Ethology* 8, 487–495.

Zuk, M., Thornhill, R., Ligon, J.D., Johnson, K., Austad, S., Ligon, S.H., Thornhill, N, and Costin, C. (1990) The role of male ornaments and courtship behavior in female choice of red junglefowl. *American Naturalist* 136, 459–473.

10 Behaviour of Horses

D. MILLS AND S. REDGATE

10.1 Origin

The ancestral wild horse was largely a social herbivore that depended on grassland, vigilance, size and speed for survival. Through domestication the horse has been successfully exploited to fulfil a wide variety of roles in society (Clutton-Brock, 1992; Hall, 2005). However, adaptability, within the constraints of good welfare, is not reflected by just physical health and long-term survival in captivity.

In the industrialized world the horse is frequently kept in social isolation, often fed a diet low in fibre, mated unnaturally and weaned at an early age. However, the ancestral phylogeny of the horse still underpins the essence of its behaviour, and failure to meet the horse's behavioural needs has led to some of the most common problems of horse management in the industrialized world. Many of these behavioural and welfare problems are associated with chronic frustration and can be managed with a greater sensitivity to the nature of the horse (Waran, 2002). However, it is equally important to recognize the ways in which a horse's behaviour can be adapted to situations which may conflict with its phylogeny. If this were not possible then the success and popularity of the horse as a domestic ridden animal would not be achievable.

There are an estimated 62 million horses worldwide distributed among 682 breeds (Hall, 2005), suggesting that the horse is one of the most differentiated of all domestic species. Its variation in size is well known, with some large percherons and shires reaching 76 inches (190 cm) at the withers in contrast to one dwarf miniature (thumbelina) standing at only 17 inches (40 cm). However, less attention has been paid to breed variation in behaviour, which is perhaps surprising given the extent of diversity. Some of the features of a breed relate to environmental factors that are independent of the domestication process; for example, animals in lower environmental temperatures tend to have bigger trunks (Bergman's Rule) and smaller extremities like ears and legs (Allen's Rule), as a result of selection for efficient heat conservation or dissipation, and it seems reasonable to assume that there might also be differences in behavioural predisposition as a result of such environmental factors.

Other features have been specifically bred for as a result of human intervention following domestication, for example the four-beat gait of the Icelandic pony (tölt) and Missouri fox trotter. In many cases the breed-typical behaviour will reflect a combination of these two selective forces (the physical environment and man), e.g. the differences in temperament often colloquially associated with blood temperature (hot blood, warmblood and cold blood). Breed differences in behaviour also extend to differences in specific learning abilities, such as discrimination learning (Mader and Price, 1980) or operant manipulation tasks (Hausberger *et al.*, 2004), and predisposition towards the development of specific behaviour disorders such as crib biting, weaving and stall walking (McGreevy *et al.*, 1995; Luescher *et al.*, 1998).

This chapter will focus on the species-typical behaviour patterns of the horse and begins with a brief review of the concept and importance of normal time budgets of the horse.

10.2 Time Budgets

The time budget of maintenance activity in horses and factors affecting it have been reviewed extensively by Houpt (2005), and so this section will highlight the importance of this concept. The majority of a horse's time in nature is spent foraging, during which the horse moves short distances frequently, sampling from one patch of herbage after another. The amount of time spent feeding will depend on the forage type and availability, but in nature it has been estimated that a horse takes around 30,000 bites per day (Mayes and Duncan, 1986) and perhaps chews nearly 60,000 times (Cuddeford, 1999).

A significant reduction in these activities – for example, by feeding a concentrate diet – may have important behavioural and physiological effects. Not only may there be an intrinsic tendency to chew, but it has been known for a long time (Colin, 1886, cited by Alexander, 1966) that saliva is produced primarily only when a horse is chewing. The volume of saliva produced and its associated mineral content is not insignificant, with a pony producing around 5–6 l/day, rich in sodium, chloride, bicarbonate and, relatively speaking, calcium too. Therefore, the level of salivation affects the level of key minerals, including bicarbonate, entering the gut, which may have knock-on effects for the gut environment, especially pH and the associated flora. This, in turn, might predispose the horse to colic and perhaps behavioural problems such as crib biting (Nicol, 1999) or wood chewing.

When grazing, a horse typically moves around foraging for between 50 and 75% of its time and spends between 15 and 35% of its time standing (Houpt, 2005). This means that the horse will typically take over 10,000 paces per day just as part of its natural grazing behaviour. Once confined the amount of movement is greatly reduced (less than 5% in a stable and around twice this value in a small paddock (Houpt, 2005)). Locomotory activity will differ between individuals, however; the measurement of activity within a stable can provide useful information as to the physical well-being and mental state of an animal. A distressed horse may be found circling repeatedly, as sudden isolation can increase locomotory activity. A change in activity level is often attributed to the horse's state of health, the diet or the management routine.

Establishing normal behavioural parameters not only is important in the identification of risk factors for potential problems with adaptation to the captive environment, but can also give insights into the early stages of other significant changes. For example, McDonnell (2005) has described the normal daily behaviour of horses in the stall (see Table 10.1), and significant deviations from this can be used to help determine chronic, acute or subacute pain.

In the following sections we consider some of the key normal behaviours of the horse in more detail.

10.3 Foraging and Feeding Behaviour

The horse is a generalist herbivore, and feral populations can be found worldwide in a variety of different habitats, ranging from savannah to desert (Boyd and Keiper,

D. Mills and S. Redgate

Table 10.1. Reference ranges for normal behavioural parameters of activity for the horse housed alone in a stall or small paddock (adapted from McDonnell, 2005).

Activity	Episodes/day	Duration
Major activity changes (eating, standing resting, standing alert, resting recumbent)	30–110	20–60 min per activity when undisturbed
Standing resting	10–30	5–120 min each; 8–12 h total
Recumbent resting	0–6	10–80 min each; 0–6 h total
Feeding on fed hay two to three times per day or continuously	10–30	5–30 min each; 4–12 h total
Drinking	2–8	10–60 s each; 1–8 min total
Urinating	4–15 (greater for mares in oestrus and stallions where marking behaviour is elicited)	
Defecating	4–15	
Rolling	2–8	2–8 rolls/bout
Erection/masturbation	18–36 (stallion) 9–24 (gelding)	

2005). Feral horses have been observed to select an annual diet of mixed plant species; in general, grass species and grass-likes are predominantly selected, along with various proportions of shrubs, leaves, stems, bark and roots (Tyler, 1972; Hansen, 1976; Lenarz, 1985; Mayes and Duncan, 1986; Putman, 1986; Gill, 1987; Duncan, 1992; Baker, 1993; Fleurance *et al.*, 2001; Menard *et al.*, 2002). Environmental conditions and plant species availability both have a heavy bearing on selection patterns. New Forest ponies, for example, will predominantly graze the grasslands during the growing seasons and will shift their selection in favour of species that provide shelter and nutrients such as gorse, holly and deciduous woodland during the non-growing season (Tyler, 1972; Putman, 1986; Gill, 1987). This flexibility towards their food selection ensures that daily intake remains high throughout the year, which is crucial to the horse's digestive strategy (Houpt, 2005).

To cope with a nutritionally variable environment the horse, like other herbivores, has adopted a patch-feeding strategy whereby preferred plant communities are regularly visited and sampled (Prache *et al.*, 1998; Fleurance *et al.*, 2001). Plants are sought and initially recognized on the basis of familiar sensory characteristics such as shape, colour, texture and flavour. Horses have been observed, for example, to select for the young, greener parts of plants that are higher in nutritional value than the rest of the available plant matter (Duncan, 1992; Menard *et al.*, 2002). These latter observations support experimental work that suggests that horses can learn to associate a food's sensory characteristics with energy- or nutrient-related post-ingestive consequences (Laut *et al.*, 1985; Cairns *et al.*, 2002; Redgate *et al.*, 2006, 2007). Conversely, horses can also learn to avoid foods that make them immediately ill (Pfister *et al.*, 2002); this association is even more persistent when the food offered is novel (Houpt *et al.*, 1990).

The usable energy gained from plants is far less than that from animal-derived protein. The horse compensates by having a digestive strategy characterized by a high voluntary intake and rate of gastrointestinal passage (Putman, 1986; Duncan *et al.*, 1990; Illius and Gordon, 1993; Sneddon and Argenzio, 1998). Thus, to achieve a high voluntary intake the horse spends a large proportion of the day feeding. Free-ranging horses spend 13–18 h/day grazing and browsing (Salter and Hudson, 1979; Duncan, 1980, 1992; Francis-Smith *et al.*, 1982; Arnold, 1984; Mayes and Duncan, 1986; Menard *et al.*, 2002). Similar total foraging times of between 15 and 18 h have been recorded with pastured horses (Francis-Smith *et al.*, 1982; Arnold, 1984). Grazing and browsing bouts occur regularly throughout each 24 h period; time of day affects bout length, as longer feeding bouts have been recorded at dawn and in the late afternoon (Tyler, 1972). Natural breaks between feeding bouts are usually short and tend to occur in order to perform other activities such as other maintenance, social and reproductive behaviours. Horses rarely fast voluntarily for more than 4 h (Davidson and Harris, 2002; Ellis and Hill, 2005), although they will take longer feeding breaks to shelter from inclement weather or biting flies (Mayes and Duncan, 1986; Gill, 1987).

Stabled horses that are provided with their total ration *ad libitum* also feed in discrete bouts, with breaks for rest or monitoring of the local environment. Feeding bouts separated by intervals of non-feeding of 10–15 min duration have been classified as meals (Ralston and Baile, 1982; Laut *et al.*, 1985). Laut *et al.* (1985) noted that, when a 10 min feeding interval was applied to 24 h observations of stalled ponies, feeding *ad libitum*, they ate approximately 17 small meals per day. Feed processing affects the duration and number of feeding bouts. Argo *et al.* (2002) demonstrated this with ponies that were presented with two feeds identical in nutritional components but presented either as pellets or as chaff *ad libitum*. The ponies had longer feeding bouts on the chaff and more frequent but shorter feeding bouts when fed the pellets.

The anatomy and function of the horse's digestive tract reflects this little-and-often or 'trickle' feeding strategy. The stomach is small and has a limited capacity to hold large amounts of food, although it rarely empties completely and digesta move rapidly on to the small intestine within 20 min of ingestion (Harris and Arkell, 2005). As there is a continuous flow of digesta from the stomach, the horse has no need for a gall bladder and bile is secreted almost continuously into the small intestine. Digesta finally move on into the large intestine; since the horse is a hindgut fermenter, a large proportion of its caloric needs are obtained through the caecal and colonic fermentation of dietary fibre (Hoffman *et al.*, 2001). The steady flow of forage through the digestive system acts as a fluid reservoir and maintains total gastrointestinal tract motility and health (Sneddon and Argenzio, 1998; Geor, 2005a).

It has been argued that the domestic horse's dietary regimen should often resemble the phylogenetic origins and natural feeding patterns of the horse (Mills and Nankervis, 1999) – i.e. with free access to species-rich grassland that has a medium to low quality energy content or *ad libitum* provision of forage and fibre-based feeds. Where horses are bred and managed specifically for leisure or sporting purposes the owner or carer has the responsibility for providing a diet that meets each individual horse's health and behavioural needs, and some compromise may be necessary. For example, the feeding strategies of racehorse trainers will be to maximize performance and, as a result, horses in training are fed high-energy feeds with little forage (Geor,

2005b), but this is not without a cost to the animal's welfare. By contrast, owners of native-type ponies often have to restrict their access to grazing and other feeds if they are susceptible to laminitis, a debilitating condition that can be triggered by the high levels of starch present in cereals and sugars present in modern horse pastures (Lockyer, 2005).

10.4 Social Behaviour

In captivity, horses are often mixed on the basis of familiarity, and little attention may be paid to the composition of the group (other than to avoid mixing stallions with others or each other); the group is frequently disrupted as one or more animals are removed for riding or other purposes. The frequency of kick injuries, as a sign of unstable social relations, should not therefore come as any surprise, but this does not indicate that animals should be kept in isolation. Horses naturally live in social groups (bands), and the benefits of social living clearly outweigh the costs for this species. These benefits would appear to relate mainly to predation avoidance, in terms of increased vigilance when grazing and safety in numbers when fleeing, and so it can be expected that there has been historically strong selective pressure for gregariousness and against isolationism. However, many modern management systems appear not to take this into account and seek to house horses alone, or at least in physical detachment from each other.

Within groups stable, long-term, linear, dominance hierarchies can be discerned among mares in single and mixed-sex groupings (Clutton-Brock *et al.*, 1976; Houpt *et al.*, 1978; Sigurjonsdottir *et al.*, 2003). Within a group, mare rank can reflect a range of factors, some of which have been found to be consistently predictive, such as size, age and length of residence within the group (Van Dierendonck *et al.*, 1995; Sigurjonsdottir *et al.*, 2003; Rho *et al.*, 2004). However, these are not totally independent, as older mares will tend to be larger and, where they are living in stable groups, the oldest will tend to have had the longest residence time. Hierarchical position is, to some extent, maintained by the outcomes of aggressive encounters involving chasing and physical shoving, threats to bite or kick and, ultimately, actual kicking and biting (Houpt and Keiper, 1982), but this is almost certainly exaggerated in the captive situation where space and the ability to retreat are more limited.

More importantly, social group living requires the formation of affiliative relationships, which are frequently reinforced and evident through mutual grooming, resting together, proximity and following behaviour and it is suggested that, in the wild, mares will typically have only one or two preferred social partners, other than their own foals, throughout their lifetime (Feh, 2005), suggesting that social relationships between mares, at least, can be very strong and enduring. This can lead to problems of isolation stress in captivity, when a mare is separated from a close social partner; however, the more general potential social stress of allowing only more transient relationships in captivity remains unexplored. Van Dierendonck (2006) suggests familiarity may be a better predictor of affiliation than kinship, and found that, among both mares and geldings, affiliates were more likely to intervene when a preferred partner engaged in affiliative behaviour such as play or mutual grooming with another individual, suggesting that horses appear to take action to protect their relationships; awareness of such behaviour may be particularly important in captive systems.

By contrast, protective and interventionist behaviour by stallions towards their breeding mares is well known. In nature a breeding stallion tends to live with a group of females (harem) and their immature offspring, isolated from other adult males (Feist and McCullough, 1975; Keiper, 1976; Miller, 1981; Berger, 1986), although variations on this theme are not uncommon – this includes breeding pairs (Feist and McCullough, 1975; Welsh, 1975; Green and Green, 1977) and the tolerance of a younger, second, stallion in certain situations to help defend the harem from other males (Miller, 1981; Berger 1986; Goodloe *et al.*, 2000). Alternatively, all-male adult bands may form (McCort, 1984). These findings suggest that stallions, too, may be more cooperative in their social relationships than is widely recognized, although a clear dominance hierarchy may also be apparent when resources are limited. All-female adult groups also occasionally arise in the wild (Goodloe *et al.*, 2000) and are common in captivity. Geldings (castrated males) tend to form their own social groups even when mares are available without an adult stallion (Sigurjonsdottir *et al.*, 2003). Geldings were also reported as playing much more than mares, with play being absent among adult mares. When resources are not limited aggression is generally rare.

Young males (up to the age of about 3 years old) are tolerated in reproductive bands until they attempt to breed with one or more mature mares, at which time the stallion will drive them out (Feh, 2005). At this point they will naturally tend to join up with males of a similar age until socially mature enough to take control of their own harem (typically around the age of 6 years). Females tend to leave the social group at about the same age when they come into oestrus. They are not usually driven out (Monard and Duncan, 1996), but they will usually refuse to mate with the familial stallion of the group and tend to seek out alternative partners at this time. Reproductive behaviour is discussed further in the next section.

10.5   Mating and Sexual Behaviour

Mares are seasonally polyoestrous, displaying repeated oestrous periods during the breeding season, with a period of anoestrus in between. The cycle is approximately 3 weeks, with 1 week of oestrus followed by a 2-week dioestrous phase, but the precise cycle length is variable between breeds and also between individuals. The oestrous period is markedly longer than for many other ungulates and, while it has been suggested (Kiley-Worthington, 1987) that this may relate to the need to establish a bond between the sexes at this time, this seems unlikely given the normal social relationship that exists between a mare and stallion. It may, however, relate to competition for the stallion between mares within the same social group (Curry *et al.*, 2007), since dominant mares may interrupt matings of subordinates or generally try to lower their reproductive fitness (Rutberg and Greenberg, 1990), and dominant mares are generally more attractive to stallions (Powell, 2000).

Sexual behaviour is a communicative process involving a complex sequence of interactions between mare and stallion, but with the mare naturally being much more proactive than is generally recognized. In semi-feral ponies, mares have been reported to initiate 88% of the pre-copulatory interactions that lead to successful mating (McDonnell, 2000). Visual cues used to attract the attention of the stallion include tail raising, which can vary in extent from a slight lift away from the body through to an almost vertical flag-like posture to expose the vulva, which may in itself act as a visual cue to the stallion, especially when combined with clitoral winking (see Fig. 10.1). The vulval labia are darkly pigmented and, when drawn back, expose the brighter pink lining of the vestibule wall. Olfactory stimuli might also be released at this time in addition to the pheromones that are present in the urine of oestrous mares. These increase interest by, and arousal in, the stallion, as demonstrated by an increased rate of flehmen response (curling of the upper lip), during which the stallion extends his neck, flares his nostrils and inhales to draw air into the richly innervated vomeronasal organ (Stahlbaum and Houpt, 1989; Fig. 10.2). A stallion will often let out a long, low nicker sound at this time.

The oestrous mare adopts a very characteristic stance, similar to that assumed for urination, with the hocks and stifles flexed, the rear limbs slightly abducted and the pelvis lowered, a position sometimes described as squatting or a sawhorse stance. In this position the mare is braced to take the stallion's weight on mounting but, in addition, the stance itself is a visual stimulus advertising the mare's oestrous status. Ostrous mares also show an increased incidence of urination, this ranging from a full evacuation to just a few drops.

If a mare is receptive a stallion will often nudge and nip along her body towards her neck. Both individuals may appear to behave quite aggressively to each other at this time but, if the mare responds favourably to a stallion's advance, his penis is extended and will become erect. He will normally make several partial mounts before attempting a full mount, usually from behind but occasionally from the side. Intromission is normally achieved after a few exploratory thrusts. Copulation does not take long and, after a few pelvic thrusts, flagging of the raised tail signals ejaculation. The head also tends to be lowered and the facial muscles relax at this time. During copulation the stallion may bite occasionally at the mare's neck and mane.

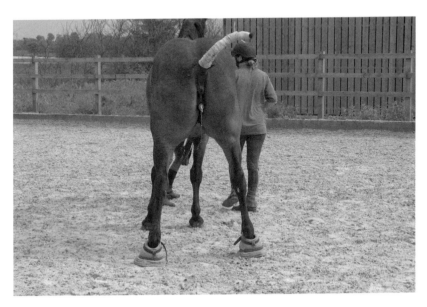

Fig. 10.1. Clitoral winking in a mare; note exposed pink vulval labia and classic wide-legged stance.

Fig. 10.2. Flehmen response in a stallion.

Within about 30 s the mare will then normally step forward to allow the stallion to dismount. It is not uncommon for a stallion to squeal a little at this time. In the natural state the stallion does not have to step backwards to dismount, although this is common in 'in hand' matings and may explain the relatively high incidence of spinal problems in breeding stallions. The stallion may be willing to serve another mare within 10 min or so, but is likely to show mated mares greater attention over the next few days. A mare in oestrus tends to be served on five to ten occasions.

These natural patterns differ markedly from what is normally allowed in the breeding of valuable bloodstock, where the stallion is brought to the mare when she is believed to be receptive, allowed to mount and then required to dismount almost immediately. Mares are usually heavily restrained at this time, to both enforce compliance and protect the stallion from damage.

10.6 Behaviour at Birth

The mare appears to be able to exert considerable control over the time of parturition and, while the normal length of pregnancy is around 340 days (Rossdale, 1967; Jeffcott, 1972), normal parturition may occur 2–3 weeks either side of this time frame. Mares will often also control the specific time of delivery to when the environment is calm and there is little apparent risk of disturbance. Foaling will therefore frequently occur at night (Jeffcott, 1972), when in the natural state the risk of predation is minimal and in captivity the risk of disturbance is also reduced. Given that the newborn is a 'follower', i.e. depends on remaining with its dam for protection, the biological advantage of this timing for such a flight-dependent prey species is self-evident. The parturition process is usually divided into three stages (Rossdale, 1967; Jeffcott, 1972).

First-stage labour: immediately before birth

At this time the mare will appear restless, wandering around a given area and will switch between one behaviour and another. She may swish her tail, turn her head towards her flanks, kick at her belly, frequently urinate, paw at any bedding, crouch, lie down and get up again. This stage may be short (minutes), prolonged (hours) or intermittent over a period of days, although the latter may be more common in captivity and result from frequent disturbance.

Second-stage labour: parturition

Bursting of the allantochorionic sac is often followed by a flehmen reaction to the uterine fluids and a nickering. The mare will normally lie on her side, with her uppermost hind leg extended. She strains forcefully and regularly to deliver the foal, but even this behaviour can be interrupted if she is disturbed; she may also pause to rest between bouts of contractions and even feed at this time. This phase usually lasts between 10 and 30 min, until the main trunk of the foal is delivered. The foal's movements usually rupture the surrounding sac and it will pull its own hind legs away from its mother. The mare should not stand immediately as there is significant blood transfer possible from the placenta to the foal at this time.

Third-stage labour: postpartum pause

Typically, the mare may remain lying for 15–20 min, but 25 per cent of foals in captivity stand within 5 min of early snapping of the cord. Such a response may result in a foal being deprived of up to a third of its blood supply, inadequate inflation of the lungs and oxygen starvation of the tissues. When the mare licks the foal, for the first time, she will often nicker and a chemical imprint is made of the foal as bonding begins. Occasionally, following a particularly difficult foaling, especially in primiparous mares, the mare may turn aggressively to the foal and reject it, before bonding occurs. The placenta is normally expelled within about 1–2 h of delivery of the foal and about two-thirds of mares will investigate it, but ingestion is rare (Virga and Houpt, 2001). Mares are naturally very protective initially of the newborn foal and will frequently display aggression towards any early approaches to their offspring. In the case of a stillborn foal, the mare will initially remain close to it but will, gradually over a period of several hours, normally lose interest in the body and resume grazing.

10.7 Care of Offspring, Including Nursing

Soon after the foal rises, the mare usually stands to allow it to suckle. She may turn to guide the foal towards the teat and will normally allow it to suckle for up to 20 min. It is suggested (Fraser, 1992) that the foal learns to suckle as a result of the successful completion of a series of modal action patterns, including coordination for standing, walking, orientation towards a dark undersurface (facilitating maternal contact), thrusting with the muzzle (resulting in teat location), sucking (to obtain milk) and suckling (to efficiently extract milk from the udder).

Inappropriate stimuli in the environment may disrupt this process and cause problems; for example, the foal may head towards a large wall-mounted hay trough in the stable instead of its mother's underbelly. The let-down of milk and suckling by the foal are both believed to play important roles in the bonding of both mother and foal, and so the artificial rearing of the orphan foal must be done with caution to avoid human imprinting. Within the first 3 h or so the foal will normally suckle again and pass the fetal faecal plug (meconium).

The foal will feed almost hourly for the first day or so. The mare will terminate about half of the nursing bouts in the first week (Tyler, 1972; Crowell-Davis, 1985) by shifting her weight from one hind limb to the other, before simply walking away from the suckling foal. Later, the mare may allow the foal to suckle as she forages. Foals are very playful and will begin solitary play within a couple of days (Boyd, 1988), and social play within the first month (Crowell-Davis et al., 1987; Boyd, 1988). The foal stays close to its mother, i.e. within about 5 m (Tyler, 1972) for virtually the whole of the first week, and is allowed to suckle at will. During this time the foal actively tries to stay near its mother when it is awake and it tends to be the mare that instigates any separation.

Mares may be aggressive to their foals if nursing is painful, e.g. if she develops mastitis or if the foal is overzealous in its suckling, and this is more likely as the foal gets older, peaking around 4–6 months of age (Barber and Crowell-Davis, 1994). Over time, the foal is increasingly responsible for any separation between the two, and the mother becomes increasingly responsible for bringing them back together (Tyler, 1972). This trend continues until the time of weaning. If the mare is pregnant, weaning will normally occur shortly before the next birth, when the foal is about 40 weeks old, otherwise it may continue to suckle for a year or more (Duncan, 1980). Mills and Nankervis (1999) suggest that, as with other domestic species, a number of sensitive phases of development can be recognized, during which the learning of certain types of association is easier for the foal, and this could be exploited more to make training of the captive horse more efficient and more effective. This is discussed further in the following section.

10.8 Offspring Development and Management

The mother's response to a range of stimuli may play a key role in shaping the behaviour of the foal (Henry et al., 2005) and this, together with the relatively immature emotional development of the young foal, has been exploited in early handling procedures such as Miller's 'imprint training' of the foal (Miller, 1991) to habituate the horse to a range of human handling procedures. Exposure of the foal to the mother in work may also be beneficial and is not uncommon in some cultures; for example, a foal may be haltered alongside its mother in harness. Clearly there is the potential to exploit this further in the captive setting, through exposing foals in a non-threatening way to a range of stimuli that it is likely to encounter in later life, a process described as 'maturation training' (Mills and Nankervis, 1999).

Weaning is, in nature, a gradual process, during which the young not only learn to be independent, but also to cope with significant frustration as they are denied access to their mother. However, in captivity, the process is usually much more abrupt and occurs much earlier, and so this process may exceed the animal's ability to adapt psychologically, with long-term consequences. Waters *et al.* (2002) have reported that weaning is associated with the onset of a range of abnormal, stereotypic behaviours, and that individual box weaning was more likely to induce these problems than group- or field-weaning strategies. They also found that the diet around the time of weaning may influence the type of stereotypic behaviour developing in later life, with the introduction of concentrates after weaning increasing the risk of crib biting; however, these problems are not simply management related, with dam social dominance and the amount of time spent suckling (Nicol and Badnell-Waters, 2005) also increasing the risk of problematic oral behaviour. While there has been growing attention to the housing and management of the adult horse in preventing welfare problems (Mills and Clarke, 2002), these findings suggest there should be equal, if not greater, attention applied to the management of the foal, as this may have greater preventive value.

References

Alexander F. (1966) A study of parotid salivation in the horse. *Journal of Physiology* 184, 646–656.
Argo, C.M., Cox, J.E., Lockyer, C. and Fuller, Z. (2002) Adaptive changes in the appetite, growth and feeding behaviour of pony mares offered ad libitum access to a complete diet in either a pelleted or chaff-based form. *Animal Science* 74, 517–528.
Arnold, G.W. (1984) Comparison of the time budgets and circadian patterns of maintenance activities in sheep, cattle and horses grouped together. *Applied Animal Behaviour Science* 13, 19–30.
Baker, S. (1993) *Survival of the Fittest. A Natural History of the Exmoor Pony.* Exmoor Books, Somerset, UK.
Barber, J.A. and Crowell-Davis, S.L. (1994) Maternal behavior of Belgian (*Equus caballus*) mares. *Applied Animal Behaviour Science* 41, 161–189.
Berger, J. (1986) *Wild Horses of the Great Basin: Social Competition and Population Size.* The University of Chicago Press, Chicago, Illinois, 326 pp.
Boyd, L.E. (1988) Ontogeny of behavior in Przewalski horses. *Applied Animal Behaviour Science* 21, 41–69.
Boyd, L.E. and Keiper, R.R. (2005) Behavioural ecology of feral horses. In: Mills, D.S. and McDonnell, S. (eds) *The Domestic Horse. The Evolution, Development and Management of its Behaviour.* Cambridge University Press, Cambridge, UK, pp. 55–82.
Cairns, M.C., Cooper, J.J., Davidson, H.P.B. and Mills, D.S. (2002) Association in horses of orosensory characteristics of foods with their post-ingestive consequences. *Animal Science* 75, 257–265.
Clutton-Brock J. (1992) *Horse Power. A History of the Horse and the Donkey in Human Societies.* Natural History Museum Publications. London.
Clutton-Brock, T.H., Greenwood, P.J. and Powell, R.P. (1976) Ranks and relationships in highland ponies and highland cows. *Zeitschrift für Tierpsychologie* 41, 202–216.
Colin G. (1886) *Traité de Physiologie Comparée*, 3rd edn. Baillière, Paris.
Crowell-Davis, S.L. (1985) Nursing behaviour and maternal aggression among Welsh ponies (*Equus caballus*). *Applied Animal Behaviour Science* 14, 11–25.
Crowell-Davis, S.L., Houpt, K.A. and Kane, L.C. (1987). Play development in Welsh pony (*Equus caballus*) foals. *Applied Animal Behaviour Science* 17, 119–131.
Cuddeford, D. (1999) Why feed fibre to the performance horse today? In: *Proceedings of the BEVA Specialist Days on Nutrition and Behaviour.* EVJ Ltd., Newmarket, UK, pp. 50–54.
Curry, M., Eady, P. and Mills, D. (2007) Reflections on mare behavior: social and sexual perspectives. *Journal of Veterinary Behavior: Clinical Applications and Research* 2, 149–157.

Davidson, N. and Harris, P. (2002) Nutrition and welfare. In: Waran, K. (ed.) *The Welfare of Horses*. Kluwer Academic Publishers, Dordrecht, the Netherlands.

Duncan, P. (1980) Time-budgets of Camargue horses. Time-budgets of adult horses and weaned sub-adults. *Behaviour* 72, 26–48.

Duncan, P. (1992) *Horses and Grasses. The Nutritional Ecology of Equids and their Impact on the Camargue*. Springer-Verlag, New York.

Duncan, P., Foose, T.J., Gordon, I.J., Gakahu, C.G., and Lloyd, M. (1990) Comparative nutrient extraction from forages by grazing bovids and equids: a test of the nutritional model of equid/bovid competition and coexistence. *Oecologia* 84, 411–418.

Ellis, A.D. and Hill, J. (2005) *Nutritional Physiology of the Horse*. Nottingham University Press, Nottingham, UK.

Feh, C. (2005) Relationships and communication in socially natural horse herds. In: Mills, D.S. and McDonnell, S. (eds) *The Domestic Horse. The Evolution, Development and Management of its Behaviour*. Cambridge University Press, Cambridge, UK, pp. 83–93.

Feist, J.D. and McCullough, D.R. (1975) Reproduction in feral horses. *Journal of Reproduction and Fertility Supplement* 23, 13–18.

Fleurance, G., Duncan, P. and Mallevaud, B. (2001) Daily intake and the selection of feeding sites by horses in heterogeneous wet grasslands. *Animal Research* 50, 149–156.

Francis-Smith, K., Carson, R.G. and Wood-Gush, D.G.M. (1982) A grazing recorder for horses – its design and use. *Applied Animal Ethology* 8, 413–424.

Fraser, A.F. (1992) *The Behaviour of the Horse*. CAB International, Wallingford, UK.

Geor, R.J. (2005a) Diet, feeding and gastrointestinal health in horses. In: Harris, P.A., Mair, T.S., Slater, J.D. and Green, R.E. (eds) *The 1st BEVA and Waltham Nutrition Symposium 'Equine Nutrition for All'*, Harrogate, UK, pp. 89–94.

Geor, R.J. (2005b) Faster, stronger, sounder: the role of nutritional supplements and feeding strategies in equine athletic performance. In: Harris, P.A., Mair, T.S., Slater, J.D. and Green, R.E. (eds) *The 1st BEVA and Waltham Nutrition Symposium*, Harrogate, UK. Equine Veterinary Journal Ltd., Newmarket, UK.

Gill, E.L. (1987) Factors affecting body condition of New Forest ponies. PhD thesis, University of Southampton, Southampton, UK.

Goodloe, R.B., Warren, R.J., Osborn, D.A. and Hall, C. (2000) Population characteristics of feral horses on Cumberland Island, Georgia and their management implications. *Journal of Wildlife Management* 64, 114–121.

Green, N.F. and Green, H.D. (1977) The wild horse population of Stone Cabin Valley, Nevada: a preliminary report. *Proceedings of the National Wild Horse Forum* 1, Reno, Nevada. Cooperative Extension Service, University of Nevada, pp. 59–65.

Hall, S.J.G., (2005) The horse in human society. In: Mills, S.D. and McDonnell, S. (eds) *The Domestic Horse. The Evolution, Development and Management of its Behaviour*. Cambridge University Press, Cambridge, UK, pp. 23–32.

Hansen, R.M. (1976) Foods of free-roaming horses in southern New Mexico. *Journal of Range Management* 29, 347.

Harris, P.A. and Arkell, K.A. (2005) How understanding the digestive process can help minimise digestive disturbances due to diet and feeding practices. In: Harris, P.A., Mair, T.S., Slater, J.D. and Green, R.E. (eds) *The 1st BEVA and Waltham Nutrition Symposium 'Equine Nutrition for All'*, Harrogate, UK, pp. 9–14.

Hausberger, M., Le Scolan, N., Bruderer, C. and Pierre, J.-S. (2004) Interplay between environmental and genetic factors in temperament/personality traits in horses (*Equus caballus*). *Journal of Comparative Psychology* 118, 434–446.

Henry, S., Hemery, D., Richard, M.-A. and Hausberger, M. (2005) Human–mare relationships and behaviour toward humans. *Applied Animal Behaviour Science* 95, 341–362.

Hoffman, R.M., Wilson, J.A., Kronfeld, D.S., Cooper, W.L., Lawrence, L.A., Sklan, D. and Harris, P.A. (2001) Hydrolyzable carbohydrates in pasture, hay, and horse feeds: direct assay and seasonal variation. *Journal of Animal Science* 79, 500–506.

Houpt, K.A. (2005) Maintenance behaviours. In: Mills, D.S. and McDonnell, S. (eds) *The Domestic Horse. The Evolution, Development and Management of its Behaviour.* Cambridge University Press, Cambridge, UK, pp. 94–109.

Houpt, K.A. and Keiper, R. (1982) The position of the stallion in the equine dominance hierarchy of feral and domestic ponies. *Journal of Animal Science* 54, 945–950.

Houpt, K.A., Law, K. and Martinisi, V. (1978) Dominance hierarchies in horses. *Applied Animal Ethology* 4, 273–283.

Houpt, K.A., Zahorik, D.M. and Swartzman-Andert, J.A. (1990) Taste aversion learning in horses. *Journal of Animal Science* 68, 2340–2344.

Illius, A.W. and Gordon, I.J. (1993) Diet selection in mammalian herbivores: constraints and tactics. In: Hughes, R.N. (ed.) *Diet Selection: an Interdisciplinary Approach to Foraging Behaviour.* Blackwell Scientific Publications, Oxford, UK, pp. 157–181.

Jeffcott, L.B. (1972) Observations on parturition in crossbred pony mares. *Equine Veterinary Journal* 4, 209–212.

Keiper, R. (1976) Social organization of feral ponies. *Proceedings of the Pennsylvania Academy of Science* 50, 69–70.

Kiley-Worthington, M. (1987) *The Behaviour of Horses.* J.A. Allen, London.

Laut, J.E., Houpt, K.A., Hintz, H.F. and Houpt, T.R. (1985) The effects of caloric dilution on meal patterns and food intake of ponies. *Physiology and Behavior* 35, 549–554.

Lenarz, M.S. (1985) Lack of diet segregation between sexes and age groups in feral horses. *Canadian Journal of Zoology* 63, 2583–2585.

Lockyer, C. (2005) How to reduce the risk of nutritionally associated laminitis. In: Harris, P.A., Mair, T.S., Slater, J.D. and Green, R.E. (eds) *The 1st BEVA and Waltham Nutrition Symposium,* Harrogate, UK. Equine Veterinary Journal Ltd., Newmarket, UK.

Luescher, U.A., McKeown, D.B. and Dean, H. (1998) A cross-sectional study on compulsive behaviour (stable vices) in horses. *Equine Veterinary Journal Supplement* 27, 14–18.

Mader, D.R. and Price, E.O. (1980) Discrimination learning in horses: effects of breed, age and social dominance. *Journal of Animal Science* 50, 962–965.

Mayes, E. and Duncan, P. (1986) Temporal patterns of feeding behaviour in free-ranging horses. *Behaviour* 96, 105–129.

McCort, W.D. (1984). Behavior of feral horses and ponies. *Journal of Animal Science* 58, 493–499.

McDonnell, S.M. (2000) Reproductive behaviour of stallions and mares: comparison of free-running and domestic in-hand breeding. *Animal Reproduction Science* 60–61, 211–219.

McDonnell, S.M. (2005) Is it psychological, physical, or both? *Proceedings of the American Association of Equine Practitioners* 51, 231–238.

McGreevy, P.D., French, N.P. and Nicol, C.J. (1995) The prevalence of abnormal behaviours in dressage, eventing and endurance horses in relation to stabling. *Veterinary Record* 137, 36–37.

Menard, C., Duncan, P., Fleurance, G., Georges, J. and Lila, M. (2002) Comparative foraging and nutrition of horses and cattle in European wetlands. *Journal of Applied Ecology* 39, 120–133.

Miller, R. (1981) Male aggression, dominance, and breeding behavior in Red Desert feral horses. *Zeitschrift für Tierpsychologie* 57, 340–351.

Miller, R.M. (1991) *Imprint Training of the New-born Foal.* Western Horseman Inc., Colorado Springs, Colorado.

Mills, D.S. and Clarke, A. (2002) Housing management and welfare. In: Waran N. (ed.) *The Welfare of the Horse.* Kluwer Academic, Dordrecht, the Netherlands, pp. 77–97.

Mills, D.S. and Nankervis, K. (1999) *Equine Behaviour: Principles and Practice.* Blackwell Science, Oxford, UK.

Monard, A.-M. and Duncan, P. (1996) Consequences of natal dispersal in female horses. *Animal Behaviour* 52, 565–579.

Nicol, C.J. (1999) Understanding equine stereotypies. *Equine Veterinary Journal Supplement* 28, 20–25.

Nicol, C.J. and Badnell-Waters, A.J. (2005) Suckling behaviour in domestic foals and the development of abnormal oral behaviour. *Animal Behaviour* 70, 21–29.

Pfister, J.A., Stegelmeier, B.L., Cheney, C.D., Ralphs, M.H. and Gardner, D.R. (2002) Conditioning taste aversions to locoweed (*Oxytropis sericea*) in horses. *Journal of Animal Science* 80, 79–83.

Powell, D.M. (2000) Evaluation of effects of contraceptive population control on behavior and the role of social dominance in female feral horses (*Equus caballus*). PhD thesis, University of Maryland, College Park, Maryland.

Prache, S., Gordon, I.J. and Rook, A.J. (1998) Foraging behaviour and diet selection in domestic herbivores. *Annals of Zootechnology* 47, 335–345.

Putman, R.J. (1986) *Grazing in Temperate Ecosystems: Large Herbivores and the Ecology of the New Forest*. Croom Helm Ltd., London.

Ralston, S.L. and Baile, C.A. (1982) Gastrointestinal stimuli in the control of feed intake in ponies. *Journal of Animal Science* 55, 243–253.

Redgate, S.E., Hall, S., Cooper, J.J., Eady, P. and Harris, P.A. (2006) Post-ingestive feedback on diet selection in horses (*Equus caballus*): dietary experience changes feeding preferences. In: *Proceedings of the International Society for Applied Ethology*, 8–12 August, Bristol, UK.

Redgate, S.E., Hall, S., Cooper, J.J., Eady, P. and Harris, P.A. (2007) Dietary experience changes feeding preferences in the domestic horse. In: *Proceedings of the 20th Equine Science Society*, 5–8 June, Maryland, 120 pp.

Rho, J.R., Srygley, R.B. and Choe, J.C. (2004) Behavioral ecology of the Jeju pony (*Equus caballus*): effects of maternal age, maternal dominance hierarchy and foal age on mare aggression. *Ecological Research* 19, 55–63.

Rossdale, P.D. (1967) Clinical studies on the newborn Thoroughbred foal I. Perinatal behavior. *British Veterinary Journal* 123, 470–481.

Rutberg, A.T. and Greenberg, S.A. (1990) Dominance, aggression frequencies and modes of aggressive competition in feral pony mares. *Animal Behaviour* 40, 322–331.

Salter, R.E. and Hudson, R.J. (1979) Feeding ecology of feral horses in western Alberta. *Journal of Range Management* 32, 221–225.

Sigurjonsdottir, H., VanDierendonck, M.C., Snorrason, S. and Thorhallsdottir, A.G. (2003) Social relationships in a group of horses without a mature stallion. *Behaviour* 140, 783–804.

Sneddon, J.C. and Argenzio, R.A. (1998) Feeding strategy and water homeostasis in equids: the role of the hindgut. *Journal of Arid Environments* 38, 493–509.

Stahlbaum, C.C. and Houpt, K.A. (1989) The role of flehmen response in the behavioural repertoire of the stallion. *Physiology and Behaviour* 45, 1207–1214.

Tyler, S.J. (1972) The behaviour and social organization of the New Forest ponies. *Animal Behaviour Monographs* 5, 85–196.

Van Dierendonck, M.C. (2006) *The Importance of Social Relationships in Horses*. Dissertation, Utrecht University, the Netherlands. Ridderprint, Ridderkerk, the Netherlands.

Van Dierendonck, M.C., De Vries, H. and Schilder, M.B.H. (1995) An analysis of dominance, its behavioural parameters and possible determinants in a herd of Icelandic horses. *Journal of Zoology* 45, 363–385.

Virga, V. and Houpt, K.A. (2001) Prevalence of placentophagia in horses. *Equine Veterinary Journal* 33, 208–210.

Waran, N.K. (2002) *The Welfare of Horses*. Kluwer Academic Publishers, Dordrecht, the Netherlands.

Waters, A.J., Nicol, C.J. and French, N.P. (2002) Factors influencing the development of stereotypic and redirected behaviours in young horses: findings of a four-year prospective epidemiological study. *Equine Veterinary Journal* 34, 572–579.

Welsh, D.A. (1975) Population, behavioural and grazing ecology of the horses of Sable Island, Nova Scotia. PhD dissertation, Dalhousie University, Halifax, Nova Scotia, Canada, 403 pp.

Behaviour of Cattle

C.B. TUCKER

11.1 Origins of Cattle

Domestication

There are two types of cattle: zebu (*Bos indicus*) and taurine (*B. taurus*). Zebu cattle have a hump at the shoulder and are found mainly in eastern Eurasia and eastern Africa. Taurine cattle do not have a hump and are the predominant cattle in the rest of the world. Cattle were domesticated between 8000 and 10,000 years ago. Until recently, it was thought that both zebu and taurine cattle were different forms originating from the same domestication event of aurochs (*B. primigenius*) or wild cattle. Recent genetic analysis suggests that the domestication process was more complex: two domestication events for taurine cattle (from *B. primigenius primigenius*), once in Eurasia, once in Africa and, a separate, third, domestication event for zebu cattle (from *B. primigenius namadicus*) (Bruford *et al.*, 2003).

Breeds

Despite the genetic and physical differences between zebu and taurine cattle, the two types can interbreed and produce fertile offspring. Zebu cattle are more tolerant of heat than taurine cattle and the two types are intermixed to create a hardy beef animal, common in hot countries like Australia. Indeed, there are hundreds of breeds of cattle throughout the world, produced through centuries of selective breeding, both within and between the two types of cattle.

In the developing world cattle serve many functions, including food production (both milk and meat), as work animals and to maintain grassland. Multi-purpose breeds or breeds adapted to local weather and grazing conditions are popular. In the industrialized world, specialized breeds dominate milk and meat production.

Specialized breeds, including Holstein-Friesian and Jersey cattle, are used for milk production. There are several common husbandry systems in the dairy industry. For example, in tie-stall barns, cows are kept indoors and tied in one location. Cows are milked, fed and lie down in their individual tie stall. Other types of barns utilize a central location for milking and cows are brought to and from the parlour. Two types of systems, free-stall barns and loose housing, utilize a central milking parlour and provide separate feeding and lying areas. Feeding and lying areas are connected by walkways and cows freely go between these resources. In free-stall barns (see Fig. 11.1a), the lying area is divided by partitions, and in loose-housing systems the lying or bedded area is open (see Figure 11.1b). Dairy cows may also be kept on pasture and bred seasonally (see Fig. 11.1c), such that calving corresponds with grass growth.

Specific breeds of cattle are used for meat production, and examples include Angus, Hereford and Charollais. Beef production is often divided into two phases:

Fig. 11.1. Common housing systems for dairy cattle include (a) free-stall barns, (b) loose housing in barns and (c) pasture.

cow–calf operations (see Fig. 11.2a) and finishing/growing to market weight (see Fig. 11.2b). Cow–calf operations breed cows and raise calves with the dam until approximately 6 months of age. Once weaned, beef cattle move to finishing operations. In North America and parts of Europe, beef cattle are typically finished on a grain-based diet in a feedlot (see Figure 11.2b). In other parts of the world, beef calves are finished on pasture.

11.2 Social Behaviour

Cattle are gregarious animals. Indeed, if separated from other animals, isolated cattle will show clear signs of stress including increased heart rate, vocalization and defecation/urination (Rushen *et al.*, 1999). Feral cattle aggregate in groups of cows and calves. Bulls form separate groups and may defend specific areas within the larger environment and intermittently interact with cow–calf groups. Humans determine group size and composition in farmed cattle, and these groups are typically either: (i) all adult or all juvenile females (dairy operations); (ii) a mix of cows, calves and a few bulls during the breeding season (cow–calf operations); or (iii) a mix of both sexes (feedlots, sometimes castrated or spayed, depending on age and practice within a country).

The structure within groups of cattle is categorized by both aggressive and affiliative behaviour. Aggressive behaviour includes threats such as lowering the head (as though to present horns) and can escalate to physical contact: head butting the head or body of another individual (see Fig. 11.3) or head-to-head pushing. In bulls, threat displays are more elaborate and include vocalizations, pawing and rubbing of the

C.B. Tucker

(a)

(b)

Fig. 11.2. Common systems for beef cattle include (a) cow–calf operations and (b) feedlots or finishing units.

Fig. 11.3. Two dairy cows at the feeding trough head butting the individual in the middle.

head on the ground and postures that make the bull look larger. The most common affiliative behaviour in cattle is allo-grooming, or social licking. Social licking between adult cattle is often directed at the neck region of the body, and cattle form grooming partnerships with specific individuals within a group.

Dominant–subordinate relationships are established and influenced by both positive and aggressive social interactions. These relationships can affect access to resources such as food, lying space, shelter and oestrous females. Aggressive interactions are common when unfamiliar individuals are mixed together, but generally decline over time as animals establish a dominance hierarchy. Aggression is also more common in intact males than in castrated steers. For example, castrated steers are less likely to initiate a fight or spar with other animals as compared with intact bulls (Price *et al.*, 2003).

Individual characteristics, such as the presence of horns and body size, can influence social success. For example, Bouissou (1972) compared social success of cattle of similar size and found that cattle with horns were dominant to dehorned cattle 85% of the time. Additionally, this same study found that, when both animals were hornless, heavier animals tended to dominate lighter animals.

11.3 Foraging and Feeding Behaviour

Cattle are both herbivores and ruminants – meaning that their natural diet consists solely of plant materials, and that their stomach is divided into four separate compartments – the reticulum, rumen, omasum and abomasum. The rumen is the largest compartment and acts as a fermentation chamber that breaks down cellulose and allows cattle to eat and digest plant material such as grasses, grains and husks. Cattle graze by gathering grass into the mouth with the tongue. By holding the grass between the bottom incisors and/or tongue against the upper palate, cattle are able to rip the grass. Cattle do not have incisors on the upper palate, and instead have a hard, ridged dental pad that is used as a grinding surface during rumination. Rumination is a distinct behaviour performed by ruminants and is sometimes called 'chewing the cud'. Cattle will ingest plant material and then, after a meal, regurgitate one bolus of partially digested food at a time, re-chew this food and then swallow again. This additional chewing helps break down cellulose or plant fibre that would otherwise be indigestible. Cattle will spend between 6 and 8 hours ruminating each day and can ruminate while either lying or standing. The length of each feeding and rumination bout varies with feed type and availability.

In addition to allowing cattle to utilize plants as a food source, the digestive tract may serve as an anti-predator strategy for ruminants. The large rumen could have allowed the ancestor of domestic cattle to ingest large amounts of food quickly while in open, perhaps risky, habitat, and then retreat to a more protected environment to ruminate and proceed with digestion.

Feeding behaviour is influenced by both the distribution and type of feed. Cows fed in a barn will typically eat for 4–6 h/day, while on pasture cattle spend more time grazing, 6–10 h/day. In addition to time spent feeding, the type of production system also affects patterns of feeding behaviour. For example, dairy cows on pasture will have more synchronous feeding behaviour and feed less at night as compared with dairy cows in free-stall barns (see Fig. 11.4a). Grass in pasture systems is spread out and all cows can feed at one time, as they do when fresh grass in provided in the morning. In free-stall barns, feed is available only at the feeding trough, and access to this may be limited by space, making it is less likely that all cows will feed at the same time.

C.B. Tucker

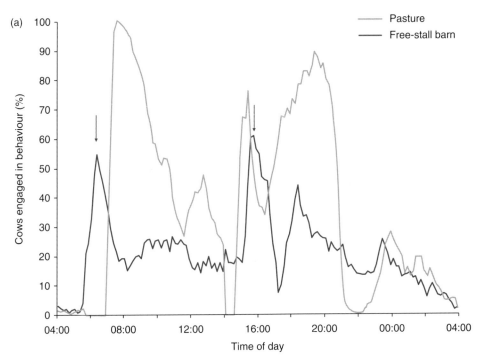

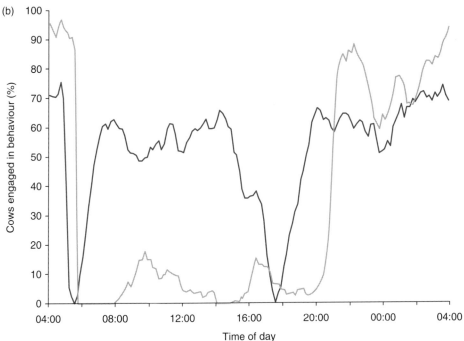

Fig. 11.4. Pattern of (a) feeding and (b) lying behaviour for dairy cattle kept on pasture (pale red line, from Tucker *et al.*, 2008) or in a free-stall barn (red line, from DeVries and von Keyserlingk, 2005). All cows were milked twice a day. The percentage of cows performing each behaviour is plotted over 24 h. The arrows in panel (a) indicate when fresh feed was delivered.

11.4 Diurnal Rhythm

Cattle show distinct diurnal rhythms when on pasture. They spend more time grazing during the day and more time lying down at night, although there is often a lying bout during the middle of the day. Indeed, cattle spend between 8 and 12 hours lying down per day, and each lying bout usually lasts just over one hour. The diurnal pattern of feeding and lying down is highly synchronized when cattle are kept on pasture. A number of housing and management factors influence these patterns.

Housing system and feed delivery

The diurnal rhythm of feeding during the day and lying down at night is less synchronous when dairy cows are kept in free-stall barns (see Fig. 11.4). Cattle kept in barns tend to spread out both feeding and lying activity throughout the day and night. However, several cues still play an important part in synchronizing their behaviour in the barn. For example, delivery of fresh feed brings cows to the feeding trough (see arrows in Fig 11.4a). On many dairy farms, feeding and milking take place at the same time. De Vries and others have separated out these two factors and found that feed delivery plays the key role in stimulating feeding behaviour. Cows increased feeding by 82% during the first hour immediately following the delivery of fresh feed, compared with a 26% decrease in feeding time during the first hour after returning to the pen from milking (DeVries and von Keyserlingk, 2005).

Weather and access to shade

Other factors, such as weather and access to shade, can affect the diurnal patterns of cattle. For example, cattle will spend more time standing and less time lying down in warm weather (e.g. Tucker et al., 2008). It is thought that cattle lose more heat when standing because air can circulate around their body. Cattle will also make other changes in order to cool down in warm weather. For example, dairy cows will spend more time standing in shade during the day and graze more at night in order to avoid the heat of the day (Kendall et al., 2006).

11.5 Mating and Sexual Behaviour

Cattle are polyoestrous, meaning they are able to breed year-round. In pasture-based dairy systems and in beef cow–calf operations, breeding is typically managed such that calves are born when grass growth begins. On barn-based dairy farms, cows calve year-round in order to produce a steady supply of milk for human consumption.

Females are usually 9 months of age at first oestrus and each oestrous cycle lasts 21 days. The duration of the receptive phases of oestrus, where the cow will stand to be mounted, lasts approximately 12 h. Receptive females (sometimes called 'on heat') will stand to be mounted by bulls, but cattle also show female–female mounting. In systems with bulls, female–female mounting can attract the attention of males. In addition to visual stimuli, bulls use olfaction to detect whether a cow is receptive.

Bulls will sniff the genital region and urine of cows and may display a flehmen response, in which the upper lip is curled while the head is outstretched. The flehmen response assists pheromones and scent molecules in reaching the vomeronasal organ in the roof of the mouth. This organ is used to detect specific chemicals typically associated with oestrus. Bulls will guard receptive females and attempt to prevent mating with other males although, depending on the sex ratio within the herd, it is likely that a female will be serviced by more than one male (Petherick, 2005).

In farming systems without males (as is the case in many dairy farms, and in some beef operations), artificial insemination (AI) is used. When a cow is artificially inseminated, previously collected semen is inserted into her reproductive track by a farm worker. Successful AI requires that humans detect the receptive phase of oestrus. There are various ways to detect oestrus in cattle, and behaviour plays a key role in each method. Cows may be observed for female–female mounting to detect oestrus. Alternatively, the tail head of the cow is painted and this paint is 'read' for signs of rubbing. If the paint wears away, this indicates that the cow has stood to be mounted and is ready to be bred. Cows also show restless behaviour during the receptive phase of oestrus. Automated measures of movement, such as a sensor on a collar or a pedometer attached to the leg, are used to record behaviour and detect the increase in activity associated with oestrus.

Bulls begin to mount conspecfics at approximately 2 months of age. By 4–6 months of age bulls will mount regularly, but do not reach sexual maturity until approximately 1–2 years of age. In systems where multiple males compete for access to females, bull age, size and dominance all affect opportunity to mate successfully with females (Petherick, 2005). Indeed, in groups of bulls, only a portion successfully sires offspring. For example, in a group of 27 bulls on a commercial cow–calf operation, five bulls sired 50% of the offspring while ten bulls sired no offspring at all (Van Eenennaam et al., 2007).

In order to assess which bulls to use for mating, libido – or both the motivation and ability to mate – is assessed with a service capacity test. Groups of bulls are placed into a pen with restrained females. Different measures of libido include time spent with each female and/or the number of mounts in a limited period – for example, 20 min (Petherick, 2005). Although libido is important for breeding success, good performance in a service capacity does not guarantee successful reproductive performance.

11.6 Parturition

Gestation in cattle lasts approximately 9 months. Both breed and sex of calf can affect gestation length. Cows separate themselves from the herd before calving, if adequate space is provided. Cows also show restless behaviour and spend more time standing the day before calving. They do not prepare a nest site, but may choose to calve indoors or at night (von Keyserlingk and Weary, 2007).

Cows give birth lying down and stand after parturition. They then begin to lick the calf. Licking serves several functions: (i) to encourage the calf to stand and suckle; (ii) to clean off amniotic fluid; and (iii) to facilitate recognition of the individual calf by the dam. Cows also engage in placentophagia, or consumption of some or the entire placenta. This behaviour may serve several functions: the placenta may have nutritional value; in addition, cattle are considered 'hider' species, i.e. cows will leave the calf behind when foraging in the first few days following parturition. Placentophagia may reduce detection of the calf by predators. In contrast, the young of 'follower' species like sheep immediately follow the dam after birth, rendering placentophagia less important for these animals.

11.7   Care of Offspring, Including Nursing

Calves usually suckle for the first time within hours of birth. Calves gain access to the udder from in front of the hind leg of the dam (see Fig. 11.2a). The first milk after birth, colostrum, contains immunoglobulins that play an important role in the development of the calf's immune system. The lining of the gut is initially porous enough to absorb these immunoglobulins, but this permeability declines rapidly in the first 24 h of life.

In cattle raised for beef, the calf remains with the cow until weaning, at approximately 6 months of age. Initially, cows seem to initiate many of the suckling bouts but, after a few weeks, calves seek out the cow. Nursing bouts are also most common early on, and decline as the calf ages. On average, calves suckle between four and ten times per day, although estimates of suckling frequency vary between studies. Initially, calves remain in peer groups or crèches for the first days of life. During this phase, cows will leave the calves to graze and will return to suckle.

Cows will also vocalize in the weeks after birth and may use vocalizations during reunification with their calf. Quiet grunting or contact calls are common immediately after parturition and decline in the first weeks of the calf's life. It is thought that these contact calls aid in auditory recognition between dam and calf (von Keyserlingk and Weary, 2007).

Occasionally, beef calves are fostered by another dam, or cross-fostered, particularly if the original dam has died. Similarly, one dairy cow may raise three or four calves from a young age. Several methods are used to establish a maternal bond with alien calves. One approach is to make the alien calf smell like the own calf by either covering it with amniotic fluid from her own calf or placing the coat of her own calf (if dead) on the alien calf as a jacket. However, this approach is viable only if the cross-fostering takes place near the time of parturition. Cows may show aggression (kicking, butting) towards an alien calf when it attempts to suckle. In order successfully to cross-foster in the days following birth, a cow may need to be restrained until she accepts the alien calf.

Although dairy calves may be raised with a foster dam, it is more common to separate them from the dam within hours of birth. Early separation of cows and calves is thought to prevent transmission of disease and improve ease of milking the cow in a milking parlour. However, calves that remain with the dam for longer periods (e.g. 2 weeks) gain more weight than their counterparts reared by humans. The behavioural response of both the cow and calf at the time of separation is more marked the more time the pair spend together (Flower and Weary, 2003).

When separated from the dam, dairy calves are reared by humans in isolation or in groups of same-age peers. Milk, milk replacer (similar to baby formula for humans) or waste milk (milk from cows treated with antibiotics that is not used for human

C.B. Tucker

consumption) is used to feed dairy calves. Regardless of the source of milk or milk product, calves are typically fed from a bucket or teat. Teat feeding systems range from simple systems with a bucket, one-way valve and rubber nipple to automated calf feeders that dispense a specified amount of food to a given individual. The amount of milk provided varies between production systems and individual farms.

Calves are highly motivated to suck. In dairy systems where calves are fed restricted amounts of milk, this motivation to suck may be unfulfilled. Cross-sucking, or sucking on another calf, typically on the ears, head and navel region, is an undesirable behaviour performed in groups. Calves are often fed at the rate of 10% of body weight during the milk-fed period and there is clear evidence that this feeding level is insufficient. There are numerous benefits associated with higher feeding levels administered through a teat, including reduced cross-sucking and much improved growth rates (De Paula Vieira et al., 2008). Finally, a combination of slower milk flow through a teat, hay feeding and access to a non-nutritive artificial teat also reduces cross-sucking, by providing more opportunity for both nutritive sucking and other forms of oral manipulation (de Passillé, 2001).

11.8   Offspring Development

Calves change how they spend their time as they age. Young calves spend a considerable amount of time lying down but, as calves get older, they spend less time recumbent. Similarly, calves with access to pasture begin to manipulate grass within the first days of life, and increase the time spent grazing as they age. Regardless of management system, all calves are dependent on milk at the beginning of their life. Their digestive system allows them to maximize the nutritional value of milk by routing the liquid directly to the abomasum or glandular stomach (similar to the stomach of monogastric animals). The oesophageal groove allows milk to bypass fermentation in the rumen. As calves age, they increase the amount of solid food and time spent grazing, such that they can survive solely on solid food.

Dairy calves are typically weaned from milk at 5–12 weeks of age. The behavioural response to weaning depends on how the process is managed. When weaning is abrupt, calves vocalize and are more active at the time milk would normally be delivered. It seems likely that calves are responding to more than simply the cessation of milk feeding: they also may take comfort in other aspects of the feeding regime. Indeed, calves will show this same behavioural response, increased vocalizations and activity, when weaned on warm water rather than milk. It is unlikely that water-fed calves are dependent on the water for nutritional reasons at this time, as they consume up to 1 kg/day of calf starter or grain. Together, these results indicate that hunger is not the only factor driving the behavioural response to weaning (Jasper et al., 2008).

Beef calves are typically weaned at 6 months of age. Abrupt weaning is common and involves physical separation of cows and calves. This process results in both nutritional and social stress, because both the dam and milk are removed at the same time. Both cows and calves show marked behavioural response to weaning: vocalization, walking and reduction in time spent eating. Calves abruptly separated from their dam will walk, on average, 16 km in one day. As with dairy calves, weaning distress is not solely nutritional. To demonstrate this point, consider an alternative to abrupt weaning, or two-step weaning. The first step in two-step weaning is to place a plastic ring in the nose of the calf. This ring prevents the calf from nursing, but allows the calf to eat solid food. Several days later the calf is separated from its mother (step two). Calves weaned in this manner vocalize and walk less, and spend more time resting compared to abruptly weaned calves (Haley *et al.*, 2005).

References

Bouissou, M.F. (1972) Influence of body weight and presence of horns on social rank in domestic cattle. *Animal Behaviour* 20, 474–477.

Bruford, M.W., Bradley, D.G. and Luikart, G. (2003) DNA markers reveal the complexity of livestock domestication. *Nature Reviews Genetics* 4, 900–910.

de Passillé, A.M. (2001) Sucking motivation and related problems in calves. *Applied Animal Behaviour Science* 72, 175–187.

De Paula Vieira, A., Guesdon, V., de Passillé, A.M., von Keyserlingk, M.A.G. and Weary, D.M. (2008) Behavioural indicators of hunger in dairy calves. *Applied Animal Behaviour Science* 109, 180–189.

DeVries, T.J. and von Keyserlingk, M.A.G. (2005) Time of feed delivery affects the feeding and lying patterns of dairy cows. *Journal of Dairy Science* 88, 625–631.

Flower, F.C. and Weary, D.M. (2003) The effects of early separation on the dairy cow and calf. *Animal Welfare* 12, 339–348.

Haley, D.B., Bailey, D.W. and Stookey, J.M. (2005) The effects of weaning beef calves in two stages on their behavior and growth rate. *Journal of Animal Science* 83, 2205–2214.

Jasper, J., Budzynska, M. and Weary, D.M. (2008) Weaning distress in dairy calves: acute behavioural responses by limit-fed calves. *Applied Animal Behaviour Science* 110, 136–143.

Kendall, P.E., Nielsen, P.P., Webster, J.R., Verkerk, G.A., Littlejohn, R.P. and Matthews, L.R. (2006) The effects of providing shade to lactating dairy cows in a temperate climate. *Livestock Science* 103, 148–157.

Petherick, J.C. (2005) A review of some factors affecting the expression of libido in beef cattle, and individual bull and herd fertility. *Applied Animal Behaviour Science* 90, 185–205.

Price, E.O., Adams, T.E., Huxsoll, C.C. and Borgwardt, R.E. (2003) Aggressive behavior is reduced in bulls actively immunized against gonadotropin-releasing hormone. *Journal of Animal Science* 81, 411–415.

Rushen, J., Boissy, A., Terlouw, E.M.C. and de Passillé, A.M.B. (1999) Opioid peptides and behavioral and physiological responses of dairy cows to social isolation in unfamiliar surroundings. *Journal of Animal Science* 77, 2918–2924.

Tucker, C.B., Rogers, A.R. and Schutz, K.E. (2008) Effect of solar radiation on dairy cattle behaviour, use of shade and body temperature in a pasture-based system. *Applied Animal Behaviour Science* 109, 141–154.

Van Eenennaam, A.L., Weaber, R.L., Drake, D.J., Penedo, M.C.T., Quaas, R.L., Garrick, D.J. and Pollak, E.J. (2007) DNA-based paternity analysis and genetic evaluation in a large, commercial cattle ranch setting. *Journal of Animal Science* 85, 3159–3169.

von Keyserlingk, M.A.G. and Weary, D.M. (2007) Maternal behavior in cattle. *Hormones and Behavior* 52, 106–113.

12 The Behaviour of Sheep and Goats

C. DWYER

12.1 Origins

Sheep and goats are ungulates (or hoofed mammals), belonging to the highly successful order Artiodactyla, the family Bovidae (including true bovines, buffalo, goats and sheep) and the tribe Caprini (comprising sheep and goats). Wild sheep and goats are among the most successful Pleistocene mammals, with wide geographical distributions extending from Europe to Siberia and Alaska to South America. Although the taxonomy of wild sheep and goats is not clear, and many subspecies exist, there are thought to be six wild species of sheep (*Ovis* genus) found in Europe, Asia and North America.

The six wild goat species (*Capra* genus) are all located in Asia and the Mediterranean basin. Although the wild ancestors of domestic sheep and goats are generally found in hill and rugged country, they are highly adaptive and have successfully colonized a variety of terrain, including desert and island habitats, and are found in the Arctic and sub-Antarctic. In Asia and Europe sheep and goats have competed for habitat, resulting in sheep occupying lower mountain slopes and hills, whereas goats are found in steep cliff areas (Fig. 12.1; Clutton-Brock, 1999). In North America the absence of competition from goats has influenced the distribution of sheep, e.g. the mountain bighorn, which range over the highest peaks. The behaviour and habitat of some of these wild sheep and goat species have been extensively studied (e.g. mountain bighorn by Geist, 1971; Asian sheep and goats by Schaller, 1977), and these studies provide valuable insights into the behaviour of the domestic animal.

History of domestication and breed development

Sheep and goats were among the first species to be domesticated, around 10,000 years ago, in the so-called 'Fertile Crescent' of the Middle East. Domestic sheep (*O. aries*) may have arisen from the domestication of more than one *Ovis* species, the mouflon (*O. orientalis*) contributing to domestic sheep in Europe and the argali (*O. ammon*) to Asiatic breeds. Domestic goats (*C. hircus*) are thought to be descended from one main species, the bezoar goat (*C. aegagrus*), although there may have been several simultaneous domestication events. The first stage of domestication is the formation of loose ties, for example by the sharing of watering places, but there is evidence of confinement and breeding control by Neolithic farmers, and the presence of 'breeds' by the Bronze Age (Ryder, 1984). Early agricultural settlements and the cultivation of crops meant that animals could be kept in enclosures at night and some protection from predators provided. In the writings of the ancient Greeks, there are descriptions of sheep herding to fresh mountain pastures in spring, and penning in winter where they were offered a range of feedstuffs (from barley, clover and lucerne to oak leaves, figs and pressed grape residues from wine making). Once

Fig. 12.1. Wild goats (*Capra ibex*) in Val d'Aosta, Italy, below Mont Blanc (image courtesy of Tony Waterhouse).

breeding came under the control of man, sheep and goat characteristics began to be shaped by man's requirements; for example, white-woolled sheep were preferred over the wild pattern of brown with a white belly. There are now in excess of 1000 sheep breeds worldwide, and nearly 600 breeds of goat.

Use and economic importance

Although sheep and goats were domesticated around the same time, sheep began to become more popular around the 1st century AD, particularly in the West. There are now 1.1 billion domestic sheep, in comparison with 700 million domestic goats, with more than three-quarters of goats found in the developing world, where there may be one goat to every three people in some countries. The ability of the goat to cope with a harsh environment and poor food quality has led to it being dubbed the 'poor man's cow'. However, both species owe their popularity to their multi-purpose ability to provide milk, meat, skins, dung for fuel and wool or fibre. They act as 'walking larders' and a visible display of wealth for nomadic peoples. In Tibet, sheep and goats are even used for portage to carry salt or grain. In India, sheep play an important role in cultural and religious rituals. Before shearing (which is considered to be a sacred communal function) sheep participate in ceremonies dedicated to Laxmi (the goddess of money), where they are washed, decorated with paint and fed jaggery (a coarse, dark sugar derived from the sap of palm trees) and coconut. Lambs born on days with important religious significance become devotional animals and neither they nor their offspring are sold or slaughtered, but confer status and respect on their owners.

C. Dwyer

Sheep and goats are generally found in regions where the climate is harsh and the terrain unsuitable for other types of agriculture. Sheep are better at coping with cold and wet climates as compared with goats, and this has aided their spread from the Middle East, with the largest sheep populations now found in China and Australia. Goats are mainly found in Africa and Asia and generally sustain small communities, in comparison with the large-scale production of wool and meat from sheep in Australia and New Zealand. Scientific research into the behaviour of sheep and goats has also favoured the sheep, although there are increasing investigations into the behaviour of domestic goats.

12.2 Social Behaviour

Both wild and domestic sheep and goats are highly social and live in small to moderate group sizes. Social living provides protection from predators, assistance in finding a mate and food, and helps with care and protection of the young. Groups are matrifocal, where females and their juvenile offspring remain together on a home range and smaller groups of males are segregated from the female flock but share an overlapping home range. Some producers of domestic sheep exploit this social group behaviour by removing the male lambs at weaning but keeping females together on an unfenced pasture or 'heft'. Cultural knowledge about the distribution of food, water and shelter sites is passed on from mothers to daughters and granddaughters, and flocks do not stray into adjoining home ranges if they are already occupied by other sheep flocks. The propensity to demonstrate home range behaviour, social group size and the strength of social attachment vary with breed, with an increase in tolerance for close contact with a larger number of animals occurring in more highly selected and domesticated breeds (Dwyer, 2004).

Following and synchronous behaviour

Mother goats and ewes form close bonds with their offspring and this encourages following behaviour, where the offspring accompany first their mother and then any adult sheep about the home range. Movement generally occurs in single file following a leader, which is an older and more independent animal. The most dominant animal in the group is usually towards the front of the movement order (as it can obtain the best grazing when arriving at the new location) but rarely leads, as this may be a more vulnerable position. The allomimetic, or synchronous, behaviours of sheep and goats may be an anti-predator strategy, as individual animals are inconspicuous when all animals are engaged in the same activity. Following behaviours and the maintenance of group cohesion are exploited by farmers when moving animals, and it is well known that moving a single animal is far more difficult than a group of sheep or goats. In many societies sheep are trained to follow the shepherd, who leads the animals to different pastures or to shelter. Some abattoirs make use of 'Judas' sheep, which live in the abattoir and will lead animals through the lairage and races to encourage easy movement.

Agonistic behaviour

Female sheep and goats do not often engage in agonistic behaviour, except when competing for limited resources, and dominance is maintained by subtle behaviours, such as eye contact and resting the chin on the back of another animal to displace them. Male animals, however, engage in impressive displays and fighting, closely linked to the morphology of their horns and skulls, primarily to gain access to females. Horns act as rank symbols, where horn size is recognized and fights do not take place between males of unequal horn size (Geist, 1971), thus reducing the risk of damage and injury. High-intensity fights, between males of equal status, are relatively rare, but can last for several hours when they do take place. The dominant feature of fights by goats is the 'clash', where opponents rear up on their hind legs and crash their head and horns together. Goat skulls are particularly thick, to protect the brain from the massive impacts that clashes cause. Although sheep also fight by clashing, only mountain sheep are reported to rear up to deliver blows, and physical fights are preceded by ritualized agonistic displays that resemble courtship behaviours (see below), including nudging, kicking with the forelegs and making 'rumbles' or growling vocalizations.

Social recognition

If two groups of sheep are mixed they will remain separate for some time, demonstrating that the sheep recognize unfamiliar animals. Over time, sheep of the same breed will become integrated into a single group; however, different breeds of sheep remain as separate groups even after prolonged exposure to one another. Recognition is based on visual and olfactory (smell) cues, and thus animals that appear visually distinct (as with two breeds) do not integrate, whereas olfactory differences diminish over time (see Fig. 12.2). Sheep have pedal scent glands on all four feet, as well as

Fig. 12.2. Social interactions between a group of recently mixed sheep following a sale. Note many sniffing interactions between animals but no overt aggression or fighting, as might be expected with other species following mixing, such as pigs (image courtesy of Fritha Langford).

C. Dwyer

inguinal and facial glands, the latter near the eye. Goats have pedal glands on only two feet and have a tail gland, which may explain why goat tails are constantly raised whereas sheep tails are usually carried low. Although scent marking is used in both species, its precise role in social recognition is unclear. The ability of sheep to distinguish familiar sheep from pictures of their faces alone has uncovered a remarkable ability of sheep to recognize more than 50 different individuals and to retain that memory for at least 2 years (Kendrick *et al.*, 2001a).

12.3   Foraging and Feeding

Sheep and goats are adapted to cope with harsh climatic conditions, and this includes an ability to utilize a wide variety of food sources. Both species lack front teeth in the upper jaw, and their smaller mouths allow them to forage closer to the ground than other ungulate species. Sheep can exploit a wide range of food sources and will eat cacti in the desert, lichens in the Arctic, tree leaves and fruit, although perhaps the most remarkable adaptation is that of the feral North Ronaldsay sheep in Orkney, UK, which forage almost exclusively on seaweed. Seaweed is structurally very different from other plants, and is low in copper but high in salt, iron and arsenic. However, the sheep are well adapted to their high salt and low copper intake and can efficiently extract copper from the seaweed, to such an extent that they suffer from copper toxicity on normal pasture. The social and foraging behaviour of these sheep is also determined by the tides, so that they are able to make best use of the forage when it is uncovered by the receding tide. Similarly, goats are known to be adventurous in their feeding habits, sometimes being used to clear areas of thick undergrowth, and they may sample non-food items such as clothes or plastic bags, if these are within reach.

Temperate sheep show another adaptation to their harsh environment by their seasonal decline in appetite, voluntary food intake and metabolic rate that occurs in winter. This appears to be an adaptation to food scarcity, as the decrease in metabolic rate precedes the decline in appetite and food intake. There is breed variation in the extent of these seasonal changes, with hill breeds of sheep showing greater decreases in winter than lowland breeds.

Digestive physiology and rumination

Sheep and goats are ruminants (as are cattle and deer), which means that they are foregut fermenters and have a stomach made up of four chambers, the rumen, reticulum, omasum and abomasum (the true stomach). The first two chambers contain microbes (bacteria, protozoa and fungi) which ferment food, including fibre. The fermented contents of the reticulo-rumen can pass between the two chambers and are regurgitated and re-chewed (known as rumination, chewing the cud or cudding). The reticulo-rumen produces volatile fatty acids, which are absorbed, and microbial protein, which, together with unfermented food residues, passes into the omasum, and then to the abomasum and small intestine. These chambers function similarly to the stomach of non-ruminant animals, with enzymatic digestion and absorption. The rumination process allows ruminant animals to obtain nutrients from fibrous plant matter more efficiently than hindgut fermenters (such as the horse).

Rumination generally occurs when the animal is lying sternally, although they will ruminate whilst standing, and typically occurs for about one-third of the day. Both species graze or forage for about 8 h/day, although this may increase to around 13 h if suitable forage is sparse. The need to find time to ruminate is a constraint on the ability of the animal to increase foraging time to maximize intake. During rumination the animal is in a state of drowsiness, and there has been speculation about whether ruminants do actually sleep. However, although the amounts of sleep and rumination are inversely related, ruminants do show periods of true sleep, often preceded by rumination.

Browsing and grazing

Although both species will eat a wide variety of foodstuffs, in general sheep are grazers that occasionally browse and goats are browsers, although they may also graze. This means that, when given the choice, sheep will preferentially eat grass and herbage, whereas goats prefer leaves and shoots from trees and bushes. Goats are well adapted to their browse diet, with mobile lips to select leaves and a digestive system that is more efficient at dealing with roughage than sheep. In addition, as agile climbers, in some environments goats will even climb into trees to obtain browse, e.g. the famous tree goats of Morocco.

Diet selection

Herbivores foraging in a variable environment (as most pastures are) are continually making decisions about which plant species, individual plants and parts of the plant to eat. The composition of their intake may vary markedly from the composition of the pasture, and comes about due to specific preferences, which may be learnt, and active avoidance of plants that are potentially toxic or may be contaminated with parasites (Forbes and Provenza, 2000). Dietary preferences show diurnal variation – for example, sheep prefer clover in the morning and grass in the evening – and seasonal variation, suggesting that animals have some memory of the location of different food types. In specific tests, sheep have been found to have excellent spatial memory and are able to learn the location of food patches after a single trial (Edwards *et al.*, 1996). Similarly, conditioned food aversions also occur after a single exposure. These abilities are clearly important adaptations for an animal foraging in a variable environment.

Many factors influence the diet selection made by sheep and goats, including breed differences, level of hunger, previous experience and social factors. Animals reared in a particular nutritional environment select a different diet from animals newly introduced to the environment, suggesting a significant learnt component to diet selection. Thus breed differences are unlikely to be due to innate differences but most likely arise due to different nutritional requirements for growth and/or the different management of breeds that will affect their previous exposure to plant types. Diet selection and foraging efficiency can also be constrained or facilitated by social factors. Ewes will not leave the social group to forage on preferred sites unless accompanied by companions, and will forage for longer on preferred sites when accompanied by familiar animals.

C. Dwyer

Water intake

Although some wild sheep and goats have colonized desert areas, both species still need to drink almost daily, although they are better adapted to coping with periods without water than other farmed animals. In general, goats are better at conserving water than sheep and, possibly due to their browse diet, may be able to obtain nearly all their water requirements from their food. In drought regions, such as parts of Australia, sheep may spend considerable amounts of time walking to water, will only forage within range of a water source and can cope with less than daily drinking only when succulent plants are available. Breed differences in ability to cope with dry environments are known to exist (Terrill and Slee, 1991), with desert breeds of sheep (fat-tailed and fat-rumped breeds) coping better than more temperate sheep breeds. The fat-tailed and fat-rumped breeds store fat in adipocytes in the rump and tail, which are more rapidly mobilized when required than back fat.

12.4 Habitat Use

Predation risks and food availability are the major evolutionary forces shaping habitat use in sheep and goats. The most important features of the habitat of wild sheep and goats are proximity to escape terrain (as flight to cover is their main anti-predator response), availability of suitable forage and a water supply. Goats and sheep are not territorial and the home range is not defended. In many situations, domestic sheep and goats are not offered a choice of habitat and may be kept in rather small, fenced paddocks. However, they still show the diurnal shifts in behaviour shown by their wild progenitors. Sheep generally camp in the hills if available, or on elevated ground, and move down to the lower regions at dawn to graze. In temperate climates, sheep graze in the morning, rest and ruminate at midday, graze again in the evening and move uphill in the evening to their campgrounds. In hot weather sheep spend more time in the shade and change their grazing patterns such that most grazing occurs in the evening and at night. Environmental complexity also influences behaviour: merino sheep kept in relatively flat and featureless terrain spend more time being vigilant than when the pasture contains more topographical variation.

Sex and age differences in habitat use

In wild sheep, adult rams will forage further from cover and for longer than ewes, are less vigilant and are more likely to be apart from the social group. Rams have reduced predation risks as compared with ewes, and are thus able to take advantage of the more abundant forage out in the open. Juvenile and young sheep are much more vulnerable to predation, are never outside the social group and always flee when a predator is present. These behaviours are reflected in the behaviour of domestic sheep, where rams are less fearful than ewes when tested in social isolation or when there is a potential predator (e.g. man) present, and young animals express greater fear than older ones.

Responses to seasons and weather patterns

Both wild and domestic sheep reduce dispersal about the home range in winter, migrate to a smaller region of the home range and decrease activity. Food preferences also show seasonal shifts, reflecting differences in availability and nutrient content of plant species and, in dry regions, the water content of plants. Sheep and goats make use of shade in hot weather, and shelter in cold and wet weather. Shelter is particularly important for goats, which have a much lower tolerance for cold, wet weather than sheep, although shelter use by sheep is affected by fleece length. Sheep, however, are more affected by hot and humid weather than goats and have a greater need for shade, preferring windy slopes in hot weather.

12.5 Mating/Sexual Behaviour

Sheep and goats are described as seasonally polyoestrous as, except for tropical breeds, they do not reproduce all year round. Reproductive behaviour is driven by the sexual activity of the female, which is influenced by day length in temperate animals (shortening days) and by rainfall in tropical goat breeds. Mating activity occurs in autumn, when males and females should be in peak condition after abundant food availability through the summer, and lambs or kids are born in spring when new plant growth occurs. As neither goats nor sheep maintain harems, mating behaviour is preceded by aggression amongst males for access to oestrous females.

Female sexual behaviour

Oestrous cycles of sheep are 17 days on average, and 21 days in goats. Oestrus and ovulation occur during a 24–48 h period when conception can occur if the ewe or doe mates. Although the sexual behaviour of the male is more overt, it is the ewe or doe which holds the initiative for mating. Oestrous behaviour is similar in both sheep and goats although it is more marked in does. The female becomes restless and vocal, bleating frequently; she may also urinate often and raise and fan her tail. Does in oestrus have been seen to mount other females, as cows do, although this is not seen in ewes. Oestrous females will seek out a male, and reproductive success is greater when the female is able to approach the male than when she is tethered. Ewes and does will show some courtship behaviour towards the preferred male, turning in front of him, rubbing along his chest and flanks and following him.

Male sexual behaviour

Males seek out oestrous females and assess their reproductive status by sniffing the anogenital area and urine, followed by the flehmen response (when the upper lip is curled, allowing access of odours to the vomeronasal organ). Male goats urinate on to their bellies, legs and beards where the scent may serve to advertise their dominance status. Once an oestrous female has been detected the male will attend and court the female. Courtship appears to be an important part of the reproductive success, at least

C. Dwyer

in wild sheep, where males that attempt to mount females without prior courtship are rejected. Courtship typically begins by the male stretching and twisting the head and neck horizontal to the ground (termed the 'low-stretch'), making low-pitched rumbling or grunting vocalizations and licking the ewe at the shoulders or flanks. The ram or buck then nudges the female with his muzzle and kicks her with his forelegs to determine whether she will stand to be mounted. If the ewe exhibits head turning (looking back towards the male) and stands then the male will mount. The male may mount between one and four times before he ejaculates. Following ejaculation there is a refractory period before the male will show interest again in another or the same female.

External factors influencing sexual behaviour

In addition to the seasonal factors that control the onset of sexual behaviour described above, sexual behaviour can be affected by nutrition, social factors and stress. Undernutrition or overnutrition, particularly of females but also of males if the decrease in food intake is severe, reduces mating behaviour and undernourished females may not exhibit oestrous behaviour. Heat stress or severe climatic conditions reduce sexual activity as both males and females seek shade or shelter, and heat stress can reduce sperm quality, resulting in poorer conception. Other forms of stress, such as fear or exposure to rough handling, can also reduce or delay oestrus in females.

The social environment can both increase and inhibit mating behaviour. The 'male effect', where the presence of the male induces oestrus in females, is known to occur in both sheep and goats, and the presence of more than one male may either enhance (through competition) or reduce (through dominance) the mating activity of other males. Early-life rearing experience is known to have a profound effect on sexual preferences, as demonstrated by a remarkable experiment where kids and lambs were cross-fostered on to ewes and does (Kendrick *et al.*, 2001b). Rams fostered on to does, or bucks on to ewes, preferentially mated with the species by which they were raised, even though they had associated with others of the same species as themselves as juveniles and adults. Although ewes and does that had been similarly cross-fostered initially also showed a preference for the species that had raised them, they changed to preferring their own species when maintained in single-species groups as adults. In a less dramatic way, the age and previous mating experience of males and females affect the completeness of their courtship behaviour, the likelihood of females seeking out males and standing, and thus reproductive success.

12.6   Maternal Behaviour at Birth

Gestation length in both sheep and goats varies with breed but generally lasts between 145 and 155 days. As with the majority of mammals, and all farmed mammals, in both species care of the young is the exclusive responsibility of the mother, and there is no paternal behaviour.

Isolation and shelter-seeking behaviour

In wild sheep and goats, the normally highly gregarious ewe or doe withdraws from the social group into remote and rugged terrain within the home range during the periparturient period. This is believed to reduce the risks of predation, and to give the mother and young the opportunity to develop a strong bond uninterrupted by other females. The extent to which this still occurs in domestic animals depends on the breed and the opportunity that females have to express this behaviour (Dwyer and Lawrence, 2005). However, even housed sheep will make some use of lambing cubicles that afford a degree of isolation. Ewes will also seek shelter at lambing time if it is cold or wet, which can markedly improve lamb survival.

Behaviour at parturition

Labour in the ewe or doe is characterized by an initial increase in restlessness and pawing, lip licking and then straining in both standing and lying positions. Neither ewes nor does build a nest in which to deliver their young. However, a birth site is selected and offspring survival is increased with increased duration spent on the birth site. Parturition is short and the lamb or kid is delivered within 1–2 h. Twinning is relatively common in domestic goats and sheep, and triplet or larger litter sizes can occur in more prolific, highly selected sheep breeds.

Onset of maternal behaviour and selectivity

Immediately after birth the mother shows intense and focused interest in the new-born, and the amniotic fluids in its coat and on the ground. This is characterized by licking the lamb or kid (see Fig. 12.3) and emitting frequent, low-pitched bleats or 'rumbles'. These behaviours serve two main purposes: first they help to dry and stimulate the lamb, and secondly the mother forms an exclusive and selective attachment to her own offspring. This attachment is initially based on the smell of the newborn lamb or kid, and mothers that cannot smell are unable to distinguish their own young from those of others. Mothers that give birth to twin or larger litters switch their grooming attention from the first lamb or kid to the newly born litter-mate, and form a new and additional olfactory attachment to that offspring, although the amount of grooming attention received by subsequent young is generally less than given to the firstborn offspring. Selective attachment is formed within an hour or less of birth, particularly in experienced mothers, and thereafter the mother will actively reject the offspring of other mothers by butting and will prevent them sucking from her. Maternal recognition of the young progresses as the lamb grows, and the mother rapidly learns to recognize her own young from a distance by voice and visual cues.

The quantity and quality of maternal care shown to the newborn at parturition can be affected by maternal experience, environmental factors, temperament and breed. Inexperienced mothers are slower to begin licking their offspring after birth and are more disturbed by the behaviour of their lamb or kid. They may butt their lamb when it moves, and move away as it tries to reach the udder, but experience with the lamb causes these behaviours to disappear and within a few hours of birth the ewe lambing for the first time will show maternal behaviour that is similar to more experienced ewes. Expression of maternal behaviour may also be reduced in ewes underfed in gestation, in ewes of more highly selected breeds compared to less selected breeds and in ewes of a nervous disposition compared to calm animals (Dwyer and Lawrence, 2005).

C. Dwyer

Fig. 12.3. Maternal licking or grooming of a newborn lamb by a Scottish blackface ewe (image courtesy of Cathy Dwyer).

12.7 Maternal Behaviour during Lactation

Suckling behaviour

Licking of the offspring gradually declines over time and, within 6 h of birth the mother no longer licks the lamb or kid, which is by then dry. Maintenance of maternal behaviour relies on behavioural cues from the neonate, particularly sucking interactions, and mothers whose offspring die lose interest in the body of their young within a few hours. During the first week of life the mother allows the offspring to access the udder and suck whenever it approaches. However, she starts to restrict access after this period by walking away when the lamb or kid tries to suck. During the first month of life the ewe will actively seek out her lamb when they are separated, and ewe–lamb distances are generally no more than a few metres. From 4 weeks onwards responsibility for approaching falls to the lamb, although the ewe still regulates the ewe–lamb distance. The ewe begins to signal to the lamb when it may approach by raising the head and bleating, the lamb may then approach the ewe and suck, although the ewe generally still ends the sucking bout. Lambs that approach when the ewe has not signalled, and twin lambs that approach without their sibling, are often not allowed to suck. Ewes or does generally do not defend their young from predators, although ewes may turn to face and threaten sheepdogs when they have young at foot, and there are reports of mountain sheep driving off avian predators with their horns (Geist, 1971).

Weaning

In the wild, lambs and kids may not be weaned until their mothers enter the rut when they are 6 months old. Management of domestic sheep and goats generally involves the removal of offspring at younger ages, around 8–12 weeks, although lambs and kids of dairy goats and sheep may be weaned within a day or so of birth. Ruminant digestion and gut development is almost complete at 8 weeks of age, so lambs or kids weaned after this age do not need to be fed milk replacer. However, although able to survive without the mother's milk supply, the psychological bond between mother and young is still present, and young will often suckle for non-nutritive, comfort reasons. Forced weaning disrupts learning and social development of the young lamb or kid and causes at least short-term stress to both partners.

In natural weaning systems, weaning seems to be related to the maternal milk supply. Sheep on better diets, because they are able to produce a greater volume of milk for longer, will wean their lambs later than ewes under poor nutrition, ewes lambing later in the season or subordinate ewes within a flock. With natural weaning, the mother increases the frequency of sucking refusal and may show agonistic behaviours towards the young.

12.8 Offspring Development

Lambs and kids are precocious and show active behavioural responses immediately after birth, standing within 30 min of birth and sucking approximately 1 h after birth. Neonates orient towards large objects and search along the maternal neck and belly to find the udder (see Fig. 12.4). Tactile and olfactory cues aid the neonate in locating the udder. Sucking behaviour both acts to provide the offspring with nutrition and facilitates recognition of the ewe by the lamb or kid (Nowak *et al.*, 1997). Both species are able to recognize their own mother at close quarters (mainly by olfactory cues) within a few hours of birth, and at a distance (by visual and vocal cues) when between 24 and 48 h old. This is important for survival for, as described above, the ewe or doe will only feed her own offspring so it is important that the offspring approach their own dam. Lambs will also associate preferentially with their twin littermates rather than with non-related lambs, from about 1 week of age, and this association is stronger if twins are of the same sex.

Ungulate offspring can be classed as either 'followers', remaining close to the mother immediately after birth, or 'hiders', where the newborn is left concealed for several hours and the mother may rejoin the social group, returning periodically to feed the young. Lambs are followers and accompany their mothers closely immediately after birth and when she rejoins the flock. Kids are usually hiders for a short period of a few days after birth, before they become followers and are closely attached to their mothers. In the hiding phase, kids lie flat and immobile until their mother approaches the location where they were left and calls to them with low-pitched bleats. Despite these different strategies, there are no species differences in the ability of lambs and kids to distinguish their dams from other ewes or does.

C. Dwyer

Fig. 12.4. Udder-seeking behaviour by a newly born merino lamb. The lamb uses tactile cues from the underside of the ewe to locate the udder, and has a reflex upward 'bunting' response to pressure on the top of the head and face (image courtesy of Raymond Nowak).

Learning species-specific behaviours

The raised head posture (or 'head-up') that the ewe uses to signal to the lamb that it may approach and suck is similar to the alarm posture used by adult sheep to communicate potential danger to the rest of the flock. By teaching her lambs to respond to this signal when young, the ewe passes on to the lamb the need to pay attention to this important signal. The flight response to potential predators also seems to have a learned component, although this may be learnt from other adults as well as the mother. By accompanying their mothers about the home range, lambs and kids also learn about shelter and water sources, and what and where food sources are. Exposure to novel foods in the presence of their mothers, even at ages when the offspring is too young to eat the food (see Fig. 12.5), facilitates acceptance of those foods later by the weaned young.

Play

Locomotor play behaviour (running and jumping) is expressed by young lambs and kids within a few hours of birth, although there are breed differences in the timing of the onset of play behaviour. The frequency of playing and complexity of play (e.g. including butting and mounting behaviours) increase with age to peak around 10–14 days of age. The lambs or kids form play bands of peers and may spend

Fig. 12.5. A 3-day-old merino lamb investigating lupin grains while its mother eats them. This behaviour facilitates learning about foods in the young lamb (image courtesy of Raymond Nowak).

periods of time distant from their mothers whilst playing. Sex differences in play behaviour emerge with males engaging in more butting and mounting play whereas females spend more time in locomotor play. Play is very infrequent after about 9 weeks of age, although play may still be expressed sporadically in juvenile and older animals.

12.9 Management and Welfare

Flocking and management

The propensity of sheep to group together and follow one another, and their flight reaction, is exploited in management where sheep can be trained to follow a shepherd, or can be herded by a shepherd and dogs. As they respond primarily to visual cues the movement of a shepherd, or the extension of an arm or crook, can cause sheep movement in the desired direction. In some societies, rather than using dogs to drive sheep, the flock follow a particular individual or the shepherd, and dogs are used primarily as guards against predators.

As the social group is very important to sheep and goats social isolation is highly stressful, and is more stressful than capture or restraint within the social group. Isolation causes agitation (running, rearing against the sides of the pen, escape attempts) and frequent, high-pitched vocalization. Allowing visual contact with other members of the social group for animals that need to be quarantined, or keeping a few 'buddy' animals alongside an animal needing special attention, can help to reduce the stress associated with isolation for management reasons.

C. Dwyer

Extensive environment and welfare

Sheep and goats are generally kept in extensive environments where, at least superficially, the domestic environment is similar to that in which the wild progenitor evolved. Thus, both species have a greater ability to express natural behaviour patterns than animals kept in more confined conditions (particularly pigs and chickens). How well farm environments meet the behavioural needs of sheep and goats has never been explored in detail; however, they do have greater behavioural freedom than most other agricultural animals. The extensive environment is not, however, without its share of welfare problems, particularly those stemming from climatic variation. Thus, extensively managed sheep and goats may be subject to temperature extremes and periods of low food and water availability. Provision of shade and shelter and access to supplementary food and water are required to keep the animals in good welfare.

Fear and predation

For sheep in particular, which may be managed in large groups with little contact with man, human contact and handling can be a significant source of fear and stress. In addition, most contacts that sheep do have with people are aversive (shearing, dipping, drenching, etc.). As sheep are very good at learning associations between places, people and negative experiences this means that most management actions, even if benign but occurring in a place where an aversive procedure has happened previously, cause sheep stress. This can be reduced to some extent by quiet, skilled and sympathetic handling and keeping aversive contacts to a minimum. Domestic goats are often kept in smaller groups with closer associations with humans, and may therefore have a more positive view of human contact, although this will still be greatly affected by the quality of handling they receive.

Extensively managed animals are more vulnerable to predation than animals managed more intensively. A number of wild carnivores will prey preferentially on sheep and goats, including foxes, bears, lynx, wolverines, wolves, coyotes, eagles, mountain lions, baboons, feral pigs and domestic dogs. Neonates and juveniles are more vulnerable to predation and the risk of predation can vary from less than 1% in countries with few large predators to nearly 30% in areas with a greater predator density. In addition to the loss of animals, predation can cause fear in the rest of flock, and attacked but surviving animals may suffer considerable injury.

References

Clutton-Brock, J. (1999) *The Natural History of Domesticated Mammals.* Cambridge University Press and British Museum (Natural History), Cambridge, UK.
Dwyer, C.M. (2004) How has the risk of predation shaped the behavioural responses of sheep to fear and distress? *Animal Welfare* 13, 269–281.
Dwyer, C.M. and Lawrence, A.B. (2005) A review of the behavioural and physiological adaptations of extensively managed breeds of sheep that favour lamb survival. *Applied Animal Behaviour Science* 92, 235–260.

Edwards, G.R., Newman, J.A., Parsons, A.J. and Krebs, J.R. (1996) The use of spatial memory by grazing animals to locate food patches in spatially heterogeneous environments: an example with sheep. *Applied Animal Behaviour Science* 50, 147–160.

Forbes, J.M. and Provenza, F.D. (2000) Integration of learning and metabolic signals into a theory of dietary choice and food intake. In: Cronjé, P.B. (ed.) *Ruminant Physiology: Digestion, Metabolism, Growth and Reproduction.* CAB International, Wallingford, UK, pp. 3–20.

Geist, V. (1971) *Mountain Sheep: a Study in Behaviour and Evolution.* University of Chicago Press, Chicago, Illinois and London.

Kendrick, K.M., da Costa, A.P., Hinton, M.R., Leigh, A.E. and Peirce, J.W. (2001a) Sheep don't forget a face. *Nature* 414, 165–166.

Kendrick, K.M., Haupt, M.A., Hinton, M.R., Broad, K.D. and Skinner, J.D. (2001b) Sex differences in the influence of mothers on the sociosexual preferences of their offspring. *Hormones and Behavior* 40, 322–338.

Nowak, R., Murphy, T.M., Lindsay, D.R., Alster, P., Andersson, R. and Uvnäs-Moberg, K. (1997) Development of a preferential relationship with the mother by the newborn lamb: importance of the sucking activity. *Physiology and Behavior* 62, 681–688.

Ryder, M.L. (1984) Sheep. In: Maon, I.L. (ed.) *Evolution of Domesticated Animals.* Longman, London and New York, pp. 63–85.

Schaller, G.B. (1977) *Mountain Monarchs: Wild Sheep and Goats of the Himalaya.* University of Chicago Press, Chicago, Illinois and London.

Terrill, C.E. and Slee, J. (1991) Breed differences in adaptation of sheep. In: Majala, K. (ed.) *Genetic Resources of Pig, Sheep and Goat.* World Animal Series B8, Elsevier Science Publishers, Amsterdam, pp. 195–233.

13 Behaviour of Pigs
M. ŠPINKA

13.1 Origins

All domestic pigs are descendants of the wild boar (*Sus scrofa*), a geographically widely distributed artiodactyl species belonging to the family Suidae. Wild boar is a versatile species capable of adjusting its way of life to different natural conditions. We will review various aspects of its natural behaviour at the beginning of each section of this chapter as a basis for understanding the behaviour of domestic pigs.

During the period when pigs started to be domesticated, the wild boar was living from northern Japan in the north-east and New Guinea in the south-east, and throughout China, tropical South Asia, India, the Middle East and up to the Atlantic coast of Europe and North Africa in the west. A marked east–west split occurred among the wild boar at latest 300,000 years ago, with 'European' populations to the west of Iran differing from those of Asia in morphology, genetics and probably behaviour.

Domestication of pigs started early in the Neolithic agricultural transition, i.e. shortly after humans switched from hunting to husbandry. Pigs were domesticated from the wild boar in at least six independent centres across its geographical range – for instance, in eastern Turkey in the 9th millennium BC, in China at about the same time and in Europe between 7000 and 5000 years BC (Giuffra *et al.*, 2000). The Chinese, Indian and European traditional pig breeds are still genetically closer to their regional wild boar counterparts than to each other. There was an introgression of Chinese pig breeds into the European stock in the 18th century. The major modern breeds arose from this mixture, some of them as late as in the mid-20th century. The interbreeding between domestic stock and the wild boar has continued in many areas throughout the history and the two forms can, of course, produce fertile offspring even today. Thus, substantial genetic variety exists within the species of *S. scrofa* in its wild, domestic and feral forms.

The domestication process exposed pigs to a different set of selection pressures from those acting upon their wild ancestors. The need effectively to escape and/or fight off predators and actively to seek diverse food sources has subsided, while the selection for large nutrient intake, efficient food utilization, fast growth and high fertility has been strengthened. However, the direction of selection has not been the same throughout the history of domestication. For instance, European domestic pigs were smaller than wild boar until the Middle Ages, while the proportion of body fat in many breeds was much larger. The recent phase of pig domestication during the last 100 years or so encompasses intense, large-scale breeding programmes for genotypes capable of turning formula-based high energy foods into lean meat quickly and efficiently. As a consequence, current production breeds of domestic pigs are much larger and heavier than the wild boar and are also capable of a higher energy turnover and of year-round reproduction.

As far as we know, no single behavioural pattern has disappeared from the wild boar repertoire during the domestication process (Gustafsson *et al.*, 1999). Therefore, behaviour of wild boar, feral pigs and domestic pigs kept in natural conditions informs us about the behavioural needs of pigs and also helps us to understand the structure and the function of behavioural patterns whose purpose is difficult to see within the barren conditions of modern intensive indoor systems. Nevertheless, the behaviour of pigs has changed quantitatively due to domestication – domestic pigs are less active, less aggressive and less wary towards potential predators (including humans), and less likely and less able actively to seek variety in their diet or food sources. It is probable that the ongoing intensive selection for production traits like large litter size, fast growth and lean body composition is modifying genetic predispositions for pig behavioural traits before our very eyes. Unfortunately, we have little understanding of these changes because the studies on the genetic basis of pig behaviour and the nature of genetic interactions between behavioural, physiological and production traits have begun only recently.

For urgent practical reasons, it is very important to study the behaviour of pigs that have been allowed to return to more natural living conditions. During the history of their lives with humans, domesticated pigs (or imported wild boar) escaped or were introduced to nature in areas previously unoccupied by wild swine. After these invasions, pigs almost invariably increased in number and spread across available land as far as it provided cover and surface water sources. The sites inhabited by feral pigs include many small and large Pacific islands, much of Australia, the south of the USA and regions in South America. Through the versatility of their feeding and activity habits and due to high fecundity, feral pigs often became serious pests because of destruction of vegetation, killing of native species, eating of crops, spreading of disease and even preying on young livestock. Control measures include hunting and even large-scale poisoning campaigns and yet often have produced only temporary effects. Sometimes, a rate of 70% annual culling is needed just to keep the populations at a stable level. Nevertheless, eradication has been achieved in several isolated areas, especially islands.

13.2 Social Behaviour

Wild and feral pigs most often live in groups of a few closely related females and their offspring from the previous year (Kaminski et al., 2005). Adult males may be associated with these herds, but more often live alone or in bachelor groups, joining the females only for breeding. However, pig groups are dynamic due to a rapid turnover of individuals and may even merge into larger associations if there are large, concentrated food resources. The home ranges of groups overlap, with no indication of between-group territoriality. Within the groups, however, a strict dominance relationship is established between each pair of animals. This dominance plays an important role in access to food, if resources are in defendable patches.

In domestic pigs, dominance relationships are less stable than in cattle. Therefore, fighting back, position reversals, harassing of subordinate animals and renewed fighting after a temporary separation can occur, even within stable groups. Nevertheless, dominance-related aggression in juvenile pigs and sows most often arises in two situations: (i) when food is available in limited space and limited time; and (ii) when groups of pigs are mixed or one or more alien individuals are added to a group. If pigs are mixed into groups larger than about 20 animals the amount of aggression per animal is lower, as many pigs refrain altogether from fighting. At the same time, a minority of animals may actually fight more intensely than in small groups (Andersen *et al.*, 2004).

Typical elements of pig fighting include sideways head knocking, pushing (shovelling), levering and biting all over the opponent's body, especially around the neck and ear region. Biting results in numerous, albeit superficial, skin lesions. The defensive manoeuvres involve turning the fore body away from the attacking pig, clinching in 'antiparallel' position and, above all, fleeing. The attacking animal usually does not pursue its fleeing opponent for more than about 3 m. Juvenile pigs mostly fight silently, while attacks by dominant sows are countered by high-pitched protest squealing of the attacked sow. Attacks at a feeding place are usually brief, since the dominant animal resumes feeding once the subordinate retreats. Nevertheless, victims of repeated food-related attacks may become unwilling to attend the feeding place. Fights during establishment of dominance can escalate into pitched battles or in the more dominant animal attacking and pursuing the other repeatedly if the latter has no possibility of escape. During the initial phases of an escalated fight both pigs attack, often circling around each other in attempts to attain a biting position. As soon as one of the animals starts using predominantly defensive manoeuvres, the outcome of the confrontation is decided (Rushen and Pajor, 1987; Fig. 13.1).

Aggression in group-housed pigs decreases welfare and production in a number of ways: (i) biting results in numerous, albeit superficial, skin lesions; (ii) aggression during post-weaning mixing contributes to other stressful aspects of weaning; and (iii) access to food or resting places is often reduced due to the fear of the defeated animals. The frequency and intensity of fighting can be ameliorated by mixing pigs

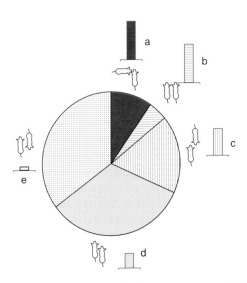

Fig. 13.1. Five different positions during fighting between unacquainted young pigs. The sections in the pie chart denote the time spent in each position, the bars depicting the biting rate (bites per time unit). In positions a–c, both pigs are attacking and the biting rate is high. However, these extremely demanding mutual attacks take up only about one-third of the time. Most of the time, either one pig is defending (the left pig in position d) or both pigs are defensive (position e) (from Rushen and Pajor, 1987).

as little as possible, designing pens so that the subordinate animals can flee and hide, and designing feeders and feeding regimes so that dominant animals cannot harass others. Both sow and weaner piglet aggressiveness, measured as the amount of aggression performed during mixing, is mildly heritable (Lovendahl *et al.*, 2005). This opens prospects for selecting against aggressiveness, especially as there seems to be a minority of extremely aggressive animals that contribute disproportionately to the overall level of fighting.

Pigs do not engage in allogrooming, and there are no reports of strong individual affiliations, although this might be due to a lack of research focus on this question. On the other hand, pigs have a strong tendency for coordination and synchronization of behaviour in space and time. For instance, one alarm bark by a pig makes the whole group, or even all pigs in a room, freeze and attend.

Humans are important components of pigs' social environment and have been so over the entire domestication period. Depending on their experience with human behaviour, pigs develop either trusting or more or less fearful attitudes towards people. Pigs can distinguish between a familiar and an unfamiliar person by relying on various cues, including the colour of their clothing. However, in most situations pigs tend to generalize their experience (either negative or positive) with specific persons to humans at large (Terlouw and Porcher, 2005). Inconsistent treatment, i.e. alternation between rough and kind handling, is perceived negatively by pigs as consistently rough handling. Thus, even limited bad experience with humans can influence pigs' attitudes towards people as a class for a long time. Therefore, it is important for caretakers to deal with pigs calmly, consistently and as positively as possible.

Educating personnel working with pigs on proper handling decreases fearfulness and is reflected in better performance of the stock. Pigs are more willing to approach a squatting or sitting person than one standing but, once the first pig in a group begins contact with the human, others quickly follow. As people are insensitive to odour signalling, and visual signals are not employed by pigs, acoustic communication is the domain through which pigs actively engage with humans. Cases from history, as well as the recent use of pigs as pets, document how an individualized human–pig relationship can develop.

13.3 Perception, Communication and Cognition

The scent sensitivity of pigs matches that of dogs. Pigs use their acute sense of scent in foraging, gathering social information and mutual communication. Wild boars locate below-ground and hidden food by smell; they can scent humans over hundreds of metres and avoid traps based on olfactory cues. In the social realm, scent is for pigs the primary way of distinguishing familiar from unfamiliar pigs, but also individuals among known pigs; pigs can recognize and remember at least 30 individuals. Other types of information extracted from the scent of the individual are its sex, reproductive status and probably dominance status.

Hearing is well developed in pigs. Lower hearing thresholds are somewhat higher in pigs than in humans for most frequencies (i.e. pigs do not hear as faint sounds as we do). Pigs' ability to locate sound sources only 5° apart is almost as good as in humans and much better than in other domestic ungulates. Unlike humans, pigs do hear some ultrasound, up to about 45 kHz.

Despite their small eyes pigs have relatively good vision, although they cannot see as accurately as humans. Similar to other ungulates, pigs have just a dichromatic vision. Pigs have a blind angle of 50–100° at the rear, thus having a wider field of vision than humans but narrower than cattle.

The pig snout disc is a powerful and yet sensitive tactile organ used in digging up and examining food and exploring features of the environment.

Pigs use primarily the olfactory and acoustic channels for communication. In the olfactory realm, adult males actively advertise their status by producing chemical compounds, two of which are the source of the unpleasant taint sometimes present in pork from non-castrated post-pubertal males: androsterone, which gives the meat a smell of urine, and skatole, smelling of faeces. Androsterone is produced in the testes and is present in saliva, urine and the preputial glands. This compound triggers the standing reflex in oestrous sows and can advance the puberty of gilts. The olfactory system of the female is five times more sensitive to androsterone than the male system. Pigs also communicate danger by the use of smell; a stressed pig releases alarm substances in its urine that can be detected by other individuals as warning signals about a danger.

Pigs have a rich repertoire of vocal signals. The pig vocal ethogram has yet to be fully mapped, but there might be up to 20 different call types. Moreover, each call can be modified quantitatively in its characteristics such as frequency, amplitude, harshness and modulation. The wide variability of calls can carry rich information about the identity of the sender, his/her location, body size, condition, motivation and emotion. Many types of pig calls (including 'normal grunting') remain poorly understood, but two areas of vocal signalling have been investigated in some depth. The first of these comprises the sows' grunting during nursing episodes, which will be described in the section on maternal behaviour, below, and the second is the 'emergency calls' that piglets emit during, for example, isolation, painful castration or when trapped under the sow's body. Calling in these situations reflects the urgency of the situation, thus inciting the mother to act or intervene. The higher the danger or need, the higher is the rate, volume and pitch of the calls (Weary and Fraser, 1995). Humans can therefore also use these calls to monitor the pigs' situation and assess their welfare. Because of the need to be heard, domestic pigs in intensive units are quite noisy. Stress-induced vocalizations are not only signs of compromised welfare but, in themselves, further negatively affect the quality of the environment both for the animals and for their carers.

Tactile communication is used by the piglets during nursing when they trigger milk ejection and communicate their hunger level to the sow through the length and intensity of teat massaging.

Wild boar use visual signals like bristle rising, ear position, tail movement and arching of the back. In the domestic pig, this communication has substantially diminished, probably due to morphological changes that prevent the signals being clearly displayed.

The pig's ability to learn from experience, memorize and combine new memories with previously available information is outstanding. This holds especially for food-related and social tasks, and thus corresponds to the fact that pigs are generalists in terms of their diet and social animals living in highly flexible social groupings. As for food acquisition, pigs can learn various associations and operational tasks, including being called to the food by individually specific acoustic signals. Pigs can deduce information about food location or preference from observing other pigs. Pigs are also able to choose options based not only on immediate rewards or punishments, but also on delayed consequences. In the social realm, pigs can discriminate not only between familiar and unfamiliar pigs, but also between individual pigs within a group or litter, not only if the 'whole' individual is present, but also on the basis of unimodal 'signatures', i.e. just the scent, the calls or the image (McLeman *et al.*, 2008). Nevertheless, pigs do not always employ their advanced cognitive abilities, but rely rather on simple cues and rules of thumb in most situations.

In spite of their tendency for synchronization and social facilitation, pigs differ a lot individually in their behaviour. Attempts to describe these differences along one dimension, e.g. passive versus active pigs, have failed. Most probably, the 'personalities' of pigs consist of several dimensions, similarly to what has been learnt about humans, other primates or dogs.

13.4   Foraging and Feeding Behaviour

Pigs are the least efficient, but most flexible, foragers among ungulates. In nature, wild boars live on a rich and varied diet. About 10% of this consists of animal food, including soil-living invertebrates, carcasses and small ground-dwelling vertebrates, but also larger, less mobile, vertebrate prey. In some areas of Australia, feral pigs kill and eat up to 40% of newborn lambs. In terms of plant matter, roots, young shoots, berries and green grasses, herb leaves and tree bark are consumed. In the longer term, however, pigs' growth and reproduction depends on a large proportion of high-energy food in their diet. This can consist of either 'mast' (acorn, beechnut), other starch or sugar-rich food (such as tree fern trunks in Hawaii) or grain crops. Accordingly, pigs strongly prefer sugary and starchy food over more fibrous diets. The current increase and expansion in wild boar and feral pig populations in many areas of the world are fuelled by crop consumption. Due to their propensity to trample crops as well as consume them, pigs can cause extensive agricultural damage (Schley and Roper, 2003). Besides energy, pigs also need protein in their diet, as they cannot synthesize ten essential amino acids. Extensive rooting of soils by wild and feral pigs is often due to their foraging for earthworms and other sources of protein. Free-ranging pigs actively move and forage for about half of their time, mostly in the mornings and evenings, although they are able to adjust their daily feeding regime to local conditions.

The modern pig breeds have been intensely selected for fast growth rate and for the ability of sows to deliver a lot of milk to the litters of up to 14 suckling piglets. For these enormous energy outputs pigs need a correspondingly massive food intake, and thus a huge appetite is now genetically encoded in current pig lines. Food mostly comes as a pre-mixed, energy-rich formula, needing no searching, foraging, extraction or mastication efforts. However, pigs need to engage in a certain amount of exploratory, foraging and feeding activity that cannot be satisfied when no bedding or other rooting and chewing material is available. This is most urgent in pregnant gilts and sows that neither grow fast nor lactate. In order to prevent them from becoming obese, these animals are kept on just 60% of their *ad libitum* intake. Although their nutritional needs are fully covered, these animals feel permanently hungry and attempt to resolve the problem through repetitive exploratory and foraging efforts, which can lead to the development of stereotypies, especially in the form of bar biting. In order to prevent stereotypies and to keep pigs occupied, the best-suited enrichment includes materials that are complex, can be manipulated or even destroyed and contain small pieces of edible matter (Studnitz *et al.*, 2007).

Pigs are flexible in the time structure of their feeding behaviour and, when accessing the feeder is difficult, they will use fewer but longer visits to achieve a similar feeding time. On the other hand, pigs have a high tendency to feed in synchrony. Group-housed growing pigs tend to eat less and grow more slowly than pigs housed individually. This can be due to competition at peak feeding times if the feeder space is limited, combined with the fact that subordinate pigs may hesitate to feed at times when all other pigs are resting. Low-ranking group-housed sows may achieve only 50–80% of the intake of their higher-ranking group mates. It is this insufficient nutrient intake, rather than the stress from the lowly position, that can negatively affect their reproduction through both lowered conception rate and reduced litter sizes.

13.5 Coping with the Environment

Wild boars spend about the same amount of time resting and being active. The active period may be concentrated mainly in daylight hours (in winter) or in mornings and evenings (in summer), or even at night (when human disturbance or hunting pressure is high). Resting occurs in well-covered places in 'daily beds'.

In domestic pigs, resting takes up to 85% of their time. Different phases of sleep (drowsiness, slow-wave sleep and REM sleep) occupy about half of the total time, with the greatest proportion of sleep occurring during darkness. Pigs prefer to rest on soft, dry, clean surfaces.

Pigs are naturally very clean animals. Piglets as young as 5 days old already defecate and urinate in places remote from their lying areas. If that is not possible and the pen is soiled, their lying time will be reduced. If the pen design allows a functional division of the available area, adult pigs use specific dunging areas for elimination. Before lying down, pigs normally check the cleanliness of the bedding and never lie in soiled areas, if they can avoid them.

For domestic pigs, the thermoneutral zone (i.e. the range of temperatures they can withstand without elevating their metabolic rate) decreases with age. Small piglets need to be kept warm, but they can secure this requirement by huddling and by digging into the bedding substrate (if such is available). In indoor housing, growing and adult pigs often have problems with overheating, but rarely with being cold. An isolated 90 kg pig on a concrete, slatted floor has a thermoneutral zone between 17 and 26°C or, without a wallowing opportunity, between 17 and 23°C. In adult pigs, activity already decreases when the temperature rises over 24°C, followed by reduced food intake and growth in even higher temperatures. Pigs do not sweat and therefore rely on behaviour to keep body temperature down in excessively warm environments (Huynh *et al.*, 2005). They cool down by lying flat on colder substrates, and especially by wallowing, i.e. by moistening the body surface in pools of water, mud or, if nothing else is available, the faeces/urine slurry (see Fig. 13.2). High respiration rate is a last-resort mechanism to dissipate heat but is in itself energy consuming. Pigs kept outdoors over the summer must be provided with shade; otherwise they may be overheated and become sunburnt. Thermal comfort can be assessed by resting behaviour: if pigs huddle intensely, they are cold; if they lie with sporadic body contact, temperature is agreeable; if all pigs lie flat on their sides and fully separated, the environment is too hot.

Fig. 13.2. Wallowing is the primary way for pigs to regulate body temperature in warm conditions.

Pigs lack allogrooming, i.e. the propensity to lick each others' body, and self-grooming is also rare. Their principal way of keeping their skin clean and free of parasites is by scratching against hard objects.

In terms of coping with the environment, the welfare of pigs is best secured if the housing method allows the animals to divide the space into areas with different functions: a dry, soft place away from the main 'traffic' for resting, feeding places with enough space for unharassed food intake, a cooler part for elimination and for foraging and an area with rooting material. Additionally, the area should be large enough for pigs to escape attacks from dominant animals. This ideal is far from reality in most current husbandry systems, but progress towards it can be made by gradually establishing minimal standards that will include elements of these requirements.

Pigs are curious animals, prepared to seize an opportunity to explore new objects and situations; this drive is stronger the more barren the environment in which they live. Housing pigs in environments that allow them to acquire new information enhances their welfare, while lack of such opportunity may even lead to underdevelopment of mental capacity (see Fig. 13.3).

When pigs become ill, they show a distinct set of behaviours, including shivering, increased sleepiness, lack of activity and reactivity, decreased food intake and vomiting. These changes support the parallel physiological and immune responses in fighting the cause of disease and reinstatement of homeostasis.

13.6 Mating Behaviour

In nature, wild boar females usually breed once a year, usually during early winter in the temperate regions of the northern hemisphere. Feral pigs can breed twice a year and spread the reproduction more evenly over the seasons if they have access to high-

M. Špinka

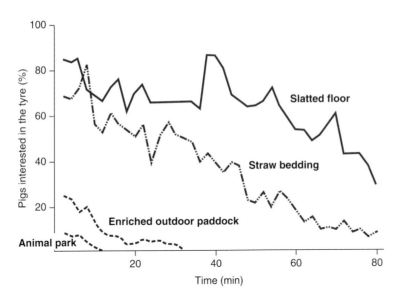

Fig. 13.3. Duration of interest in a new object (a tyre) by pigs kept in four different environments. Pigs in barren indoor environments do not have their need to explore satisfied, and therefore show a rebound behaviour by exploration of a new object (redrawn after Stolba and Wood-Gush, 1981).

energy foods. With up to four piglets per litter surviving until weaning, wild and feral pig population can quickly recover after trapping, poisoning or hunting actions aimed at reduction of the population.

Oestrus is highly synchronized in wild boar females within a group, but not necessarily between groups. Domestic sows usually give birth to about two litters per year, with very little seasonal variation. In intensive systems, where piglets are weaned between 2 and 6 weeks of age, sows usually come into oestrus about 5 days after being separated from their young. The so-called standing oestrus (the time during which the sow will stand still in response to boar stimulation) lasts 1–3 days and, about two-thirds of the way into this period, ovulation occurs. As the fertility of shed eggs declines within 4–8 h, and the fertile life of boar sperm in the reproductive tract is longer (about 24 h), the optimal time for mating or insemination is sometime before ovulation, i.e. about 12 h into the first day of oestrus; this can then be repeated every 12 h. If a sow fails to conceive, new oestrus periods follow every 18–24 days. When sows are allowed to suckle their piglets for many weeks in the presence of a boar (for instance, in the so-called family systems), they readily show lactational oestrus and therefore can become pregnant while still nursing.

About 2 days before the start of standing oestrus, sows begin to show proceptive behaviour, i.e. active displays of the upcoming receptivity (Pedersen, 2007). This proceptive behaviour includes spending time near the boar or his pen, nosing the flanks of other sows and mounting. Dominant sows can mount subordinate females, but mounting 'up the hierarchy' is actively repelled until the period of standing oestrus of the dominant animal. Proceptive behaviours peak in frequency when the receptive behaviour starts. Receptive behaviour consists of the sow reacting to the boar's smell and grunting calls by stopping movement, possibly by squealing and urinating and, if the boar mounts, by standing still under his weight. This standing reaction is also used for manual checking of whether a sow is in oestrus. The test is much more reliable in the presence of the boar, or at least with a simultaneous application of the boar pheromone.

In the wild, male pigs mark their presence to females by leaving saliva marks on trunks and possibly through more frequent wallowing. The sexual behaviour of the domestic male pig starts with courting females that behave proceptively (Hemsworth and Tilbrook, 2007). During courtship, boars approach females, utter short series of characteristic grunts, champ their jaws while salivating and may urinate rhythmically. If the female stands firm, the boar continues with sniffing her head and anogenital region, nosing or nudging her flanks, and eventually mounting. Ejaculation lasts many minutes. A boar can copulate up to eight times over a period of several hours. A sow may mate with more than one boar during one oestrous period. and thus litters of mixed paternity may occur.

Boars differ individually in both their sexual motivation (willingness to mate) and their sexual dexterity (ability to achieve intromission and ejaculation). Isolation rearing, high air temperature and presence of dominant males all decrease sexual performance, while observation of other males mating may enhance it.

13.7 Periparturient Behaviour

The potential reproductive rate of the wild boar is the highest among all ungulates. This high fecundity level has been taken to the extreme during domestication and modern breeding. This is inherently connected with relatively little concern about individual piglets and therefore high mortality risks for the young, especially during the period around birth.

Free-ranging sows separate themselves from their group before parturition and seek a partly sheltered place in which to build a nest. Wild boar females usually nest in dense cover, in warm places and near water. If the ambient temperature is low, free-ranging pigs build large, well-insulated nests, whereas in hot temperatures nesting is reduced to bedded, shallow hollows in the ground. Nest building starts about 15 h before parturition under the influence of increased prostaglandin levels (see also Chapter 3). The sow first walks around, sniffs and roots at the ground or the floor and thereafter switches to carrying and arranging of the available nest material. If no suitable material is available, the nest-building attempts continue after parturition has begun.

During the parturition itself and immediately following, sows normally remain lying passively, with their teats exposed, under the influence of endorphins. Piglets are born at intervals of about 15 min; they immediately struggle to the udder and wander from teat to teat, sampling the colostrum which is rich in energy and immunoglobulins. Early colostrum intake is vital for newborn piglets, since they have limited body reserves and their immune system is not yet functional at birth. In large litters, the last-born piglets may suffer from reduced colostrum intake because most of it has been consumed by their earlier-born siblings. The sow sniffs the piglets as they pass her snout and learns to recognize their identity within the first day, and thereafter rejects any other piglets as alien. Therefore, cross-fostering of piglets between litters should be done as soon as possible after parturition.

M. Špinka

Within hours, each piglet starts to develop an attachment to a specific teat. The piglet then defends its teat vigorously against its siblings by using sharp canines and incisors capable of inflicting slashes on its opponents' faces. In large litters, the difficulty of securing a teat against competition from stronger siblings may decrease the growth and survival chances of weaker piglets. In less numerous litters, some piglets may use two teats and this creates the opportunity for later incursion of alien piglets into group-housing systems.

During parturition and the first 2–3 days thereafter the average cumulative piglet mortality is about 15%, even in well-managed herds. The two main causes of death are undernutrition and crushing by the sow. Very often these two factors combine, as small and hungry piglets stay closer to the sow and are thus more frequently crushed. Crushings occur when a sow lies down or rolls over. When a piglet becomes trapped under the sow, it starts squealing. If the sow reacts to the squealing and wriggling of the piglet by standing up within a minute the piglet usually survives, but longer crushing periods are mostly fatal. Therefore, the early ability of the sows to satiate piglets with colostrum/milk, her careful changes in position and a strong response to piglet screams are characteristics that enhance piglet survival. Plans to select for these behavioural traits are being developed in spite of the fact that their heritabilities are low (Gade *et al.*, 2008).

13.8   Behaviour during Lactation

Immediately after birth milk is available almost continuously, but quickly a pattern emerges of synchronized suckling every 50 min or so (Drake *et al.*, 2008). Each nursing episode begins with 1–2 min of teat massage by the piglets. This massaging triggers oxytocin release from the sow's pituitary gland, this hormone causing milk ejection that lasts only 20 s. This is the only period when milk is available and it can be recognized by fast, rhythmical sucking movements by all piglets in the litter. Thereafter, piglets switch back to teat massaging and continue with it for several minutes. Nursing is accompanied by rhythmical grunting of the sow that serves to call the piglets to the udder and to announce – through an increase in the grunting rate – the upcoming milk ejection (see Fig. 13.4). Sows in auditory contact also use grunting to synchronize their nursings closely in time. About 10–35% of all nursing episodes do not contain milk ejection and piglets acquire no milk during these. When lactating sows are kept in groups, suckling of piglets that do not belong to the same litter is quite common, but does not present a major problem in stabilized groups.

During the first 2 weeks of lactation, the initiative in suckling gradually shifts from the sow to the piglets (Drake *et al.*, 2008). Initially, the sow starts almost all nursings by exposing her udder in a lateral lying position and also allows the piglets to massage the mammary glands for many minutes after each milk ejection. Gradually, the sow leaves it more often to the piglets to initiate the nursings by starting to nudge her udder and thus stimulating her to assume the nursing position. On the other hand, the sow gradually restricts the duration of post-ejection teat massage and may later even start nursing in the standing position. In this way both the frequency of nursing and opportunities for the piglets to massage the udder decrease. There seem to be two behavioural feedback mechanisms between piglets' milk intake and milk

Fig. 13.4. Suckling piglets ingest milk synchronously during the brief periods of milk ejection. The approaching milk ejection phase is announced by the sow vocally, and therefore piglets listen attentively in order not to miss the milk flow.

production by the sow. Hungry piglets engage in more intense and longer final teat massage, which may, in turn, increase the milk output on following nursings. Second, hungry piglets attempt to initiate nursing episodes at shorter intervals, thereby increasing the milk output per episode.

In free-ranging pigs, sows stay around the nest (and the piglets mostly within the nest) for a couple of days after parturition, whereafter they join the group and mix with its members. On most intensive pig farms, lactating sows are confined in a narrow metal crate. Although the crate may somewhat reduce the mortality of piglets due to crushing, it hinders nest building and, later in lactation, prevents the sow from leaving the piglets and thus escaping their sucking attempts. Sows housed in individual pens during parturition and lactation generally enjoy better welfare. Outdoor farrowing huts are a widely used option in some countries. Group housing of lactating sows can work well under careful management, especially if combined with individual housing for parturition and the first week of lactation.

13.9 Piglet Behavioural Development

Wild boar females and domestic sows in semi-natural enclosures wean their litters at about 4 months of age. In contrast, piglets are usually weaned between 3 and 5 weeks of age on commercial farms, although either earlier or later weaning is also practised in some countries. The most common way of weaning is to separate piglets from their mother, move them to another room or building, mix several litters together and switch them abruptly to solid food. This procedure exposes piglets to several stressors at an age when they are ill-prepared to withstand them (Weary *et al.*, 2008). Weaning is therefore invariably connected with growth check and often also with increased frequency of diarrhoea. Piglets weaned at 3 weeks or earlier often engage in belly nosing, an abnormal nudging of the bellies of their peers (see Fig. 13.5). Weaning can be made easier for piglets by weaning later, by removing the sow and by keeping piglets in their home environment, and also by providing solid food prior to weaning. Post-weaning litter mixing results in intense fighting. Masking odours, tranquillizers, dim light and maternal pheromones have all been tried for reducing fighting, with mixed results. When piglets from different litters have the opportunity to mingle with each other before 3 weeks of age, they become much more tolerant of alien piglets after weaning.

Piglets play by the first day of life. The frequency of play behaviour increases until about 3 weeks of age, whereafter it declines, although even adult sows can be stimulated to play by fresh straw. Play is usually quite synchronized within litters, comprising varied sequences of solitary locomotory elements like scampering, dashing, pivoting, flopping on the belly and on the side and head tossing, combined with patterns of play fighting (often performed in awkward positions, e.g. while sitting) and play chasing. Play is important as a welfare indicator since it is performed only when animals are not hungry, ill, stressed or fearful. Play also enhances welfare through the immediate self-rewarding effect, and probably also through enhancement of later coping abilities. Play is stimulated by moderately challenging stimuli such as new bedding material, access to a new space or meeting alien piglets (before 3 weeks of age).

Rearing pigs in barren environments that discourage natural foraging, sow–piglet interaction and social interactions such as play renders them less able to withstand

Fig. 13.5. Early-weaned piglets tend to develop a specific type of abnormal behaviour called belly nosing.

stresses such as weaning, being manipulated by humans, mixing and transportation. Minimal improvements, such as adding a little straw, do not alleviate the situation. However, alternatives – such as a more spacious pen with enough bedding and some enrichment objects – have been proved to make a long-term difference for the development of pigs.

Acknowledgements

Writing of this chapter was supported by grant MZE0002701404 from the Czech Ministry of Agriculture.

References

Andersen, I.L., Naevdal, E., Bakken, M. and Boe, K.E. (2004) Aggression and group size in domesticated pigs, *Sus scrofa*: 'when the winner takes it all and the loser is standing small'. *Animal Behaviour* 68, 965–975.

Drake, A., Fraser, D. and Weary, D.M. (2008) Parent–offspring resource allocation in domestic pigs. *Behavioral Ecology and Sociobiology* 62, 309–319.

Gade, S., Bennewitz, J., Kirchner, K., Looft, H., Knap, P.W., Thaller, G. and Kalm, E. (2008) Genetic parameters for maternal behaviour traits in sows. *Livestock Science* 114, 31–41.

Giuffra, E., Kijas, J.M.H., Amarger, V., Carlborg, O., Jeon, J.T. and Andersson, L. (2000) The origin of the domestic pig: independent domestication and subsequent introgression. *Genetics* 154, 1785–1791.

Gustafsson, M., Jensen, P., De Jonge, F.H., Illmann, G. and Špinka, M. (1999) Maternal behaviour of domestic sows and crosses between domestic sows and wild boar. *Applied Animal Behaviour Science* 65, 29–42.

Hemsworth, P.H. and Tilbrook, A.J. (2007) Sexual behavior of male pigs. *Hormones and Behavior* 52, 39–44.

Huynh, T.T.T., Aarnink, A.A., Gerrits, W.J.J., Heetkamp, M.J.H., Canh, T.T., Spoolder, H.A. M., Kemp, B. and Verstegen, M.W.A. (2005) Thermal behaviour of growing pigs in response to high temperature and humidity. *Applied Animal Behaviour Science* 91, 1–16.

Kaminski, G., Brandt, S., Baubet, E. and Baudoin, C. (2005) Life-history patterns in female wild boars (*Sus scrofa*): mother–daughter postweaning associations. *Canadian Journal of Zoology* (*Revue Canadienne de Zoologie*) 83, 474–480.

Lovendahl, P., Damgaard, L.H., Nielsen, B.L., Thodberg, K., Su, G.S. and Rydhmer, L. (2005) Aggressive behaviour of sows at mixing and maternal behaviour are heritable and genetically correlated traits. *Livestock Production Science* 93, 73–85.

McLeman, M.A., Mendl, M.T., Jones, R.B. and Wathes, C.M. (2008) Social discrimination of familiar conspecifics by juvenile pigs, *Sus scrofa*: development of a non-invasive method to study the transmission of unimodal and bimodal cues between live stimuli. *Applied Animal Behaviour Science* 115, 123–137.

Pedersen, L.J. (2007) Sexual behaviour in female pigs. *Hormones and Behavior* 52, 64–69.

Rushen, J. and Pajor, E. (1987) Offence and defence in fights between young pigs (*Sus scrofa*). *Aggressive Behaviour* 13, 329–346.

Schley, L. and Roper, T.J. (2003) Diet of wild boar *Sus scrofa* in Western Europe, with particular reference to consumption of agricultural crops. *Mammal Review* 33, 43–56.

Stolba, A. and Wood-Gush, D.G.M. (1981) The assessment of behavioral needs of pigs under free-range and confined conditions. *Applied Animal Ethology* 7, 388–389.

Studnitz, M., Jensen, M.B. and Pedersen, L.J. (2007) Why do pigs root and in what will they root? A review on the exploratory behaviour of pigs in relation to environmental enrichment. *Applied Animal Behaviour Science* 107, 183–197.

Terlouw, E.M.C. and Porcher, J. (2005) Repeated handling of pigs during rearing. I. Refusal of contact by the handler and reactivity to familiar and unfamiliar humans. *Journal of Animal Science* 83, 1653–1663.

Weary, D.M. and Fraser, D. (1995) Calling by domestic piglets: reliable signals of need? *Animal Behaviour* 50, 1047–1055.

Weary, D.M., Jasper, J. and Hötzel, M.J. (2008) Understanding weaning distress. *Applied Animal Behaviour Science* 110, 24–41.

14 Behaviour of Dogs

D.L. WELLS

14.1 Origins and Domestication

The domestic dog, *Canis familiaris*, is one of more than 36 species of canid, a family belonging to the order Carnivora and including, amongst others, the wolves, foxes, coyotes and jackals. The canids are a widely distributed group of terrestrial predators, which, with some exceptions, are relatively lithe and powerful in build and social in nature.

Many theories have been proposed over the years to explain the ancestry of the domestic dog (for reviews see Coppinger and Coppinger, 2001; Jensen, 2007; Miklósi, 2007). These range from the idea that the dog arose from a wild type of canid somewhat similar to today's Australian dingo or New Guinea singing dog, to the notion that the dog has multiple ancestors, including, for example, the coyote, *C. latrans,* and golden jackal, *C. aureus*. Genetic evidence now suggests that the wolf, and perhaps more specifically the grey wolf, *C. lupus* (see Fig. 14.1), is the closest living relative of the domestic dog; whether or not the modern-day wolf is the true ancestor of the dog, however, is still questionable and very much under scrutiny.

There is also great debate as to where and when the process of domestication actually took place. The fossil record is fairly consistent in suggesting that the dog was domesticated around the end of the last Ice Age, somewhere between 12,000 and 14,000 years ago (Davis and Valla, 1978). More recent studies of molecular genetic markers, however, have yielded contrasting results. Thus, Vilà and colleagues (1997) estimated that the dog may have diverged from multiple wolf matrilines anywhere between 100,000 and 135,000 years ago. Subsequent work, however, has pointed to a single gene pool of ancestry some 15,000–40,000 years ago, with the origins of this pool lying somewhere in East Asia. More recent research has questioned the conclusions of this earlier work and indicated that, in line with the palaeontological records, 15,000 years ago is the most likely time for the divergence of domestic dogs from wolves, with multiple, independent domestication events having taken place over the years. The exact number of domestication events for dogs still remains unknown, although the recent unravelling of the dog genome (Lindblad-Toh *et al.*, 2005) may help to shed more light on this in the years to come.

Domestication, and further artificial selection by humans, has given rise to a wide variety of dog breeds. Indeed, today, there are over 400 breeds of domestic dogs, with newly recognized cross-breeds (e.g. Labradoodles, Schnoodles, Cockapoos) being created and gaining popularity all the time. Many of these breeds show enormous morphological diversity. Some (e.g. German shepherd dog, Japanese Akita, Siberian husky) have retained their wolf-like appearance. Others (e.g. poodle, chihuahua, bichon frise), however, have been highly neotenized and now bear little resemblance to their wild progenitor (see Fig. 14.2). The functions of dogs in today's society are as diverse as their physical appearance. While most are kept exclusively for the

Fig. 14.1. The grey wolf, *Canis lupus*, is the closest living relative of the domestic dog.

Fig. 14.2. The Siberian husky (a), Labrador retriever (b) and the bichon frise (c), although physically different, are all descended from the wolf.

purpose of companionship, many serve as working animals and are involved in, amongst other duties, the herding of sheep, guarding of property, detection of criminals, drugs and explosives and the provision of assistance for those with disabilities (for review see Coppinger and Coppinger, 2001).

14.2 Social Behaviour

By and large, the Canidae are social carnivores, with a propensity to develop and dwell in groups. The composition of these groups varies significantly across both species and individual populations. Some attempts have been made to categorize the canids according to their social organization and behaviour. Michael Fox (1975), for example, has suggested that these animals can be loosely divided into three types. Type 1 includes solitary hunters such as the red fox, *Vulpes vulpes*, that generally develop temporary pair bonds during the breeding season. The male may stay to assist with the rearing of the pups, but will usually disappear once the offspring can fend for themselves. Type 2 canids include those species that develop more of a long-standing pair bond, such as the dingo or coyote, *C. latrans*. An abundance of food may keep the unit together, although offspring will usually disperse if and when food supplies become scarce. Type 3 canids include the wolf. These animals generally live in packs of between two and 30 members, with group size being regulated by the availability of food and other resources (Mech, 1970). Packs normally develop complex dominance hierarchies, or 'pecking orders', which play an important role in the determination of access to privileges and initiatives including travelling, hunting and reproduction. While normally monogamous, with animals developing long-term pair bonds, polyandrous matings in wolves have been witnessed.

The social behaviour of domestic dogs is as diverse as that of their wild cousins. Feral dogs provide an interesting link between the social behaviour of the wolf and the pet dog. True feral dogs (as opposed to free-ranging animals that may have owners or associate themselves with a particular household) generally dwell in groups of between two and six animals. These are not as tightly governed by pecking orders or efforts to develop cohesiveness, however, as the average wolf pack. Although wolf packs are comprised of genetically related members, feral dog groups are not typically related and group membership may be more transient in nature. Unless socialized with humans early on in their development (see later), feral dogs may show fearful reactions towards people, making great efforts to avoid them.

Pet dogs, which spend much of their time in the company of humans, develop less of a pack mentality than their feral counterparts or wild ancestors; none the less, there is some suggestion that many pet dogs come to regard their owners as members of the pack and their house as the 'den'. Unlike wolves, pet dogs can learn how to integrate themselves into the family unit, and in most cases develop complex social relationships with their human caregivers (see Fig. 14.3). The nature of these relationships is very much owner–dog specific, and the success of the pairing is dependent upon the types of bond that are established. Inappropriate bonding (e.g. over-attachments, under-attachments), coupled with a lack of obedience training, can sometimes lead to the development of psychological problems, which manifest themselves through the animals' behaviour, e.g. separation anxiety, dominance aggression. Some of these

D.L. Wells

Fig. 14.3. The domestic dog is one of the most popular companion animals in Western society, and for many people an integral part of family life.

problems can be rectified by behavioural modification (for examples of common behaviour problems and appropriate treatments see Overall, 1997); many, however, culminate in the abandonment of dogs on to the streets, or relinquishment to animal rescue shelters.

14.3   Foraging and Feeding Behaviour

The Canidae are opportunistic scavengers that will ingest a wide range of animal and vegetative matter. Much of the diet consumed depends upon the species of canid under scrutiny and the geography of the habitat. Thus, North American grey wolves generally live on elk, caribou and deer during the winter months, consuming rodents and lagomorphs in the summer.

Other wild canids may feed less exclusively on prey, and subsist on a wider range of available food types. Some jackals, for instance, have been noted as living primarily on carrion, supplemented by fruit, grass and insects such as dung beetles. Coyotes, likewise, have been shown to eat a diverse range of foods including rabbits, hares, rodents, plants and insects. Arctic foxes, *Alopex lagopus*, which live in a relatively harsh environment, eat almost any form of meat readily available. Prey typically consists of lemmings and voles, but can also include ground squirrels, seal pups, bird eggs and remnant scraps from polar bears. This species, and other canids that hunt alone, often hoard surplus food in caches, in some cases large enough to feed an individual for up to a month. Domestic dogs can often be witnessed doing the same thing with excess food or other resources, e.g. toys. Here, the front paws are used to dig a hole, the mouth or nose is used to place the relevant item inside, and the nose is employed to push over the earth and conceal the resource.

Feral dogs, like wild canids, are opportunistic scavengers. Again, much of their diet depends upon the availability of food items. City-dwelling feral dogs often forage from bins and waste sites and may consume a wide variety of foods, much of which has been prepared for human consumption. Some feral dogs will kill and consume prey specific to the area. Several authors, for instance, have witnessed feral dogs on the Galapagos Islands feeding on marine iguanas, *Amblyrhynchus cristatus*, and others in Venezuela's Llanos preying on capybara, *Hydrochaerus hydrochaeris*. Some concern has recently been expressed over wildlife depletion arising from feral dog predation on livestock populations. Scientific studies of such animals in North America and Italy, however, have revealed little, if anything, in the way of such activity, and much of the damage to wildlife is now thought to be due to free-ranging and stray dog populations.

The feeding behaviour of pet dogs is very much under the control of humans. A diverse range of dry (generally cereal-based) and wet (meat-based) foods are available on the market for the pet dog population. Although diet is ultimately controlled by the owner, and may be related to factors including price, smell and ease of preparation, dogs often show preferences for certain types of food over others. Studies have shown that dogs prefer cooked over raw meat, meat over cereal-based foods and show a ranked preference for beef over pork, followed by lamb, chicken and finally horsemeat (for a review see Bradshaw and Thorne, 1992). Like humans, some dogs can develop intolerances to certain components of food, giving rise to physical and, in some cases (and still awaiting more scientific evidence), psychological problems; a wide range of speciality meals (e.g. wheat-free, preservative-free) are now available on the open market to cater for these animals.

In addition to the daily diet provided by their owners, some pet dogs, like their wild cousins, scavenge opportunistically. It is not unusual to see some dogs, notably those bred for hunting purposes (e.g. terriers), stalking and killing small rodents. Others may ingest faeces (coprophagy), belonging to either themselves or another animal. The aetiology of this behaviour pattern in dogs remains largely unknown, although several anecdotal suggestions have been proposed. Some believe that stool-eating indicates an unbalanced diet or pancreatic enzyme deficiency; faeces are thus ingested in an attempt to gain more nutrients. Others adhere to the belief that coprophagy in dogs results from inappropriate operant conditioning, arising from either the attention directed towards the dog by its owner, or the actual taste of the faeces.

14.4 Communication

Dogs communicate with both their conspecifics and humans using a wide variety of sensory systems, including vision, olfaction and audition.

Vision

In relation to some animals (e.g. primates, cats), dogs are not a particularly vision-driven species. Indeed, it has been estimated that the dog's eye for detail is approximately six times poorer than the average human's. Colour perception is also relatively limited in this species. Although not entirely colour-blind, as often assumed, dogs are believed to perceive the world in the much the same way as a red/green colour-blind person. From an evolutionary point of view, the dog has not needed these types of visual characteristic to survive; instead, however, it has developed a good ability to see in dim light and detect subtle movements and motion.

Dogs use a range of visual signals to communicate their intentions to others (see Fig. 14.4). The positioning of the body, ears, tail and direction of eye gaze can all convey important information about status and behavioural intention (Abrantes, 2003). Dominant animals are more likely to exhibit an upright body posture, erect ears and tail and maintain direct eye contact. In some cases, an aggressive dominant individual may bare its teeth and raise its hackles, i.e. the fur on its back (pilo-erection). All of these visual cues are designed to make the animal look bigger, more threatening and, ultimately, reduce the need for physical combat. Subordinate individuals, by contrast, generally make themselves look smaller, thus lowering the body, ears and tail and breaking eye contact earlier than their more dominant counterparts. Extremely submissive animals may roll over on their back, present their inguinal region and, in some cases, even urinate. Aggressive–submissive animals may display a rather confusing mixture of visual signals that, on the one hand, imply confidence and a desire to fight (pilo-erection, bared teeth) but, on the other, convey a desire to avoid the situation and take flight (flattened ears, low body posture, low tail).

All of the above visual signals can be conveyed by dogs to other animals and humans. The latter, however, are not always adept at accurately interpreting many of the more subtle behavioural cues, and this lack of understanding can, in some unfortunate situations, culminate in bites and other injuries.

Fig. 14.4. Body posture can convey important information about status and behavioural intention. The dog on the left, with its flattened ears and lowered body, is showing signs of submission; the dog on the right, with its upright tail and firm stance, is showing signs of dominance.

Although dogs can and do employ visual signals to communicate their intentions, the highly neotenized appearance of some breeds, with their floppy ears, docked tails and longer fur, has led to a reduced ability accurately to deliver and/or interpret visual information. Domestic dogs are thus probably more reliant on olfactory signals as a method of communication.

Olfaction

The domestic dog is well renowned for its remarkable sense of smell, with the olfactory acuity of this species lying somewhere between 10,000 and 100,000 times better than that of humans. Using its sense of smell, the dog can discriminate between the odours of its own species, other species (e.g. humans) and successfully match odours to sample. The dog's remarkable sense of smell, and ability to be trained easily, has resulted in it becoming widely employed by organizations around the world for purposes including tracking, search and rescue, cadaver recovery and explosive, narcotic, mould and gas detection. More recently, it has been shown that dogs can even be trained to use their sense of smell to detect underlying health problems, including certain types of cancer (for a review see Wells, 2007).

Olfaction also plays an important role in canid communication. Three types of olfactory signal are utilized for this purpose – the deposition of *urine, glandular secretions* and *faeces*. Urine is the most commonly employed olfactory cue, the composition of which can convey information about individual status and may be utilized to denote territories and home ranges. The action of urination can also communicate useful visual information. Male dogs in particular (although not exclusively) generally show raised-leg urination (RLU). Sometimes males will cock their legs without the production of urine – a so-called raised-leg display (RLD). This is particularly common in the presence of other dogs, suggesting a visually driven display function of the raised leg.

Glandular secretions are also used by dogs for the purpose of communication. There is a particularly high concentration of sudoriferous glands – which produce watery secretions – around the head and anal regions. Dogs are prone to sniffing these areas, suggesting that the secretions play a role in olfactory communication. The release of apocrine and sebaceous secretions from the anal glands during defecation also provides key cues, and may be employed in individual and territorial recognition.

Faecal matter is the least likely product of elimination to be utilized by domestic dogs for communication; the wolf, by contrast, has been noted to defecate at the junctions of trails, presumably to convey information on territorial boundaries. Some domestic dogs will scratch the ground with their hind legs after defecation (and sometimes urination). The precise function of this behaviour pattern is still unclear. One explanation is that the scratching helps to spread the scent of the faecal material; in most cases, however, the deposition is seldom touched, weakening the plausibility of this theory. Other types of odour cue, however, may still be spread via the scratching. The dog's sweat glands are located on the pads of its paws, and the sebaceous glands are positioned in between the toes; scratching may therefore help to spread these specific types of glandular secretion. The action of scratching the ground may also provide visual information to conspecifics, in much the same way as the RLD discussed earlier.

D.L. Wells

Audition

Dogs have a remarkable sense of hearing. Not only can they perceive sounds across a wide spectrum of frequencies, but they can often pinpoint the exact source of a sound and accurately discriminate between a variety of auditory signals. Dependent upon breed and age, the average domestic dog can hear sounds in the spectrum of 67–45,000 Hz; this compares with a frequency range of 64–23,000 Hz in humans. The mobility of the dog's ear, which, in many breeds, can prick up and move around, in combination with the 'cupped' shape, explains, in part, the dog's acute sense of hearing.

Auditory signals in the form of vocalizations are often used as a method of communication. The domestic dog can elicit a wide variety of vocalizations, including grunts, growls, whines, yelps, coughs and, of course, barks. The wolf seldom barks, with this type of signal comprising only about 2% of lupine vocalizations. Moreover, the bark tends to be context-specific in the wolf, used mainly as an alert or challenging threat. The domestic dog, in contrast, barks very frequently, for extremely lengthy periods of time and in virtually every behavioural context.

The divergence in bark behaviour between dogs and wolves has prompted researchers to suggest that barking in domestic dogs is generally applicable and may lack specific communication functions. Recently, however, it has been shown that bark structure in dogs actually varies predictably with context (Yin, 2002). Barking, and other dog vocalizations, may thus be a more complex form of communication than previously thought. Research will hopefully shed more light on this in the years to come.

14.5 Biological Rhythms

The wild Canidae vary substantially in their biological rhythms. Wolves and foxes are generally active at night, whilst dingoes and African wild dogs, *Lycaon pictus*, show a crepuscular pattern of diurnal behaviour, with peak periods of activity occurring at dawn and dusk, when temperatures are cooler and the opportunities for hunting prey are greater.

The domestic pet dog is diurnal in nature, tending to sleep when its owner does. However, unlike humans, the dog is a polyphasic sleeper, having several bouts of sleep per night. The average sleep episode lasts for just under 1.5 h, with mean waking periods being around 40 min. Most dogs also sleep several times during the day; the frequency and duration of these sleeping bouts vary substantially between animals, being dependent upon individual dog–owner routines. As in humans, the canid sleep cycle comprises both periods of quiet (i.e. non-REM) and rapid eye movement (REM) sleep. It is during this latter stage of sleep, in which animals are frequently witnessed to paddle their legs and utter vocalizations, that dogs are believed to dream. Dogs may be able to shift to nocturnal activities quite easily. Studies of urban stray dogs, for instance, have shown that such animals can be quite active during the night and early morning.

Sleep disturbances can appear with increasing age, with older animals sleeping more during the day, and less at night. This is often a feature of the degenerative disorder cognitive dysfunction syndrome (CDS), a condition that has been likened to Alzheimer's in humans.

14.6 Sexual Behaviour

The domestic dog is sexually mature by about 7–8 months of age; this is in stark contrast to the wolf, which does not reach sexual maturity until approximately 22 months. Male dogs are capable of producing sperm, and hence reproducing, at any time of the year, although most females are dioestrous, coming into season ('heat') only twice annually; one exception is the basenji, a breed that originates from central Africa, and only comes into season once a year. The average female first comes into season between 6 and 12 months of age.

The oestrous cycle of the bitch comprises four stages – *pro-oestrus*, *oestrus*, *metoestrus* (or *diestrus*) and *anoestrus*. Pro-oestrus lasts for up to 2 weeks and is characterized by a bloody discharge, restless behaviour and increasing attractiveness to male animals. Most females are unreceptive to mating during this stage of the cycle, although males may attempt mounting.

During oestrus, the female is receptive to mating. This typically involves sniffing of the anogenital region by the male, mounting, penetration and ejaculation of sperm. In most cases, the penis is held for some time inside the vagina, and the animals are consequently locked together in a tie position, often facing opposite directions. Ovulation normally takes place, even in the absence of a mating, on day two of the oestrus phase. Oestrus may last for anywhere between 10 and 21 days, the length of which is dependent upon whether or not the bitch has been mated.

The metoestrus period lasts for around 2 months, and may constitute the period of pregnancy if mating, and successful fertilization of the egg, has occurred. Pseudo-, or false, pregnancies are relatively common in the dog, with some females producing milk, building nests and displaying other signs of maternal behaviour. These behaviour patterns tend to disappear once hormone levels have stabilized. In rare cases, some bitches may have a false pregnancy with each heat.

The final stage of the oestrous cycle, anoestrus, represents the period of reproductive inactivity, and lasts for an average of 4 to 5 months.

14.7 Behaviour at Birth and Parental Care

The gestational period of the domestic dog is approximately 63 days. One or two days prior to parturition, the bitch may become restless, seek seclusion and shred papers, blankets or bedding in an attempt to build a 'nest'. Whelping (parturition) very often takes place overnight, and lasts anywhere between 3 and 6h. The first signs of labour are panting and vaginal discharge. The first puppy normally takes the longest to whelp, with an average interval of 45 min between the birth of individuals. Many whelpings proceed without complication or human intervention; however, some births may need to be closely monitored, and Caesarean sections are not uncommon, particularly amongst breeds with large heads.

The average litter size is six to seven, although larger breeds tend to have considerably more (and relatively smaller) puppies than smaller breeds. Puppies are born in individual placental sacs containing amniotic fluid; these are normally removed by the mother using her teeth and sometimes ingested, providing a useful source of nutrients. Puppies are licked by the mother immediately after birth for the purpose of cleaning and to stimulate breathing. Most puppies, guided by both olfactory cues and

Fig. 14.5. Rottweiler puppies suckling from their mother. Puppies suckle exclusively for the first 3–4 weeks of their lives, following which weaning begins.

maternal nudging, attach themselves to a teat and start suckling within the first hour of birth (see Fig. 14.5). Mothers often continue to clean their pups whilst suckling, particularly the anogenital area, in a bid to stimulate elimination. This procedure is repeated for the first few days after birth, as puppies are unable to urinate or defecate without the aid of their mother's tongue.

The mother focuses almost exclusively on the cleaning and sucking of her puppies for the first 2–3 weeks of life, following which she starts to leave them alone for maybe an hour or two a day. Puppies suckle from their mother for the first 3–4 weeks of life, following which weaning begins and episodes of nursing decrease. Around this time, the mother may start to push puppies away from her teats and encourage independence. Most puppies are fully weaned by 8 weeks of age.

14.8   Development of Behaviour

The behavioural development of the domestic dog can be divided into five stages: *prenatal, neonatal, transitional, socialization* and *juvenile*.

Prenatal period

The prenatal phase covers the entire period from conception to parturition, and lasts approximately 63 days. This period of development is still relatively shrouded in mystery. It has recently been discovered, however, that the domestic dog, like many other species, can learn about chemosensory cues (i.e. odours and/or tastes) during the prenatal period, and that such learning may play an important role in shaping olfactory-guided development and behaviour following birth (Wells and Hepper, 2006).

Neonatal period (0–14 days)

The canid puppy is relatively helpless for the first 2 weeks of its life, relying heavily upon its mother for suckling, elimination and comfort. Most of the puppy's time is dominated by feeding and sleeping. The eyes and ears are closed and non-functional during this stage, with the animal relying most heavily upon its senses of smell, taste and touch. Whilst the puppy can move, motor abilities are limited and often reflex-driven. Vocalizations, including whines, grunts and mews, are relatively common, and often used as a method of alerting or attention-seeking.

Transitional period (14–21 days)

As the name suggests, the transitional phase constitutes a period of rapid neurological and physical change, the most apparent of which is the opening of the eyes around day 13, and ear canals between days 18 and 20. The puppy is no longer reliant upon its mother for the stimulation of elimination, and can both urinate and defecate of its own accord, usually outside the nest. Motor movements are more advanced than during the preceding neonatal period, with the animal being able to crawl forwards and backwards, stand up on four paws and walk, albeit clumsily. During the transitional period, puppies start to interact with one another, displaying their first signs of play fighting and tail wagging. Vocalizations become louder and are produced in response to a wider variety of stimuli.

Socialization period (3–10 weeks)

This relatively long period of development sees the onset of more adult patterns of behaviour. The pup's sensory and motor abilities are fully developed, allowing it to explore and learn about its surroundings. The puppy gradually spends more and more time away from its mother, relying less on feeding and sleeping. Teeth begin to erupt at the start of the socialization period, and weaning is normally complete by 7–8 weeks of age.

   The socialization phase was once considered to be a *critical* period in the development of social relationships (Scott and Fuller, 1965). Nowadays, it is more generally recognized as a *sensitive* period of development, during which the pup develops important social bonds with its mother, litter-mates and humans. Inappropriate socialization during this period can sometimes lead to fear-induced behavioural problems that often persist into adulthood, and in many cases are resistant to modification.

Juvenile period (10 weeks to sexual maturity)

The juvenile period of development runs from approximately 10 weeks of age up to sexual maturity. The pup grows rapidly during this stage, with most breeds reaching their full height and weight by about 8 months. Adult teeth replace milk (temporary) teeth at roughly 5 months of age. Most puppies are removed from their mothers at the start of this period, and continue to develop social relationships with humans and other animals. General behaviour is similar to that expressed by puppies during the socialization phase, but is more advanced and controlled in nature. Male dogs start to show raised-leg urination (see earlier), although the age of onset of this varies between individuals. Puberty is a rather gradual process in male dogs, with animals expressing an interest in females that are in oestrus as early as 4 months of age, but not normally attempting to mate until 7–8 months; puberty in females, by contrast, is much more sudden, occurring at the time of the bitch's first season, anywhere between 6 and 12 months of age.

References

Abrantes, R.A. (2003) *The Evolution of Canine Social Behaviour.* Wakan Tanka Publishers, Naperville, Illinois.

Bradshaw, J.W.S. and Thorne, C.J. (1992). Feeding behaviour. In: Thorne, C.J. (ed.) *The Waltham Book of Dog and Cat Behaviour.* Pergamon Press, Oxford, UK, pp. 115–129.

Coppinger, R. and Coppinger, L. (2001) *Dogs: a Startling New Understanding of Canine Origin, Behavior and Evolution.* Scribner, London.

Davis, S.J.M. and Valla, F.R. (1978) Evidence for domestication of the dog 12,000 years ago in the Natufian of Israel. *Nature* 276, 608–610.

Fox, M.W. (1975) *The Wild Canids: their Systematics, Behavioural Ecology and Evolution.* Van Nostrand Reinhold, New York.

Jensen, P. (2007) *The Behavioural Biology of Dogs.* CAB International, Wallingford, UK.

Lindblad-Toh, K., Wade, C.M., Mikkelsen, T.S. *et al.* (2005) Genome sequence, comparative analysis and haplotype structure of the domestic dog. *Nature* 438, 803–819.

Mech, L.D. (1970). *The Wolf: the Ecology and Behaviour of an Endangered Species.* Natural History Press, New York.

Miklósi, A. (2007) *Dog Behaviour Evolution and Cognition.* Oxford University Press, Oxford, UK.

Overall, K.L. (1997) *Clinical Behavioral Medicine for Small Animals.* Mosby, London.

Scott, J.P. and Fuller, J.L. (1965) *Genetics and the Social Behaviour of the Dog.* University of Chicago Press, Chicago, Illinois.

Vilà, C., Savolainen, P., Maldonado, J.E. *et al.* (1997) Multiple and ancient origins of the domestic dog. *Science* 276, 1687–1689.

Wells, D.L. (2007) Domestic dogs and human health: an overview. *British Journal of Health Psychology* 12, 145–156.

Wells, D.L. and Hepper, P.G. (2006) Prenatal olfactory learning in the domestic dog. *Animal Behaviour* 72, 681–686.

Yin, S. (2002) A new perspective on barking in dogs (*Canis familiaris*). *Journal of Comparative Psychology* 116, 189–193.

Behaviour of Cats

J. BRADSHAW

15.1 Origins and Domestication

The cat was the first species to be domesticated solely for the purpose of controlling pests, namely the rodents exploiting the stores of dried grain that followed the domestication of cereal crops (Diamond, 1997). However, precisely where this occurred and when are still unclear: the first unequivocal evidence that we have for fully domesticated cats is later, when the cat had attained religious significance in Egypt and was being bred in temples.

The ancestral species is the wild cat, *Felis silvestris*, which is still widely distributed, though often in marginal habitats, from northern Europe to southern Africa and western India. The DNA of modern domestic cats suggests that they first became isolated from their wild ancestors somewhere in the Fertile Crescent, which runs from Turkey to Egypt, supporting the theory that they were domesticated for their ability to control mammalian pests. The precise location remains unclear; the most recent study (Driscoll *et al.*, 2007) did not sample any cats from North Africa, and an earlier study (Randi and Ragni, 1991) identified the North African/Arabian subspecies *F. silvestris lybica* as the most likely candidate. The choice of *F. silvestris* appears to have been fortuitous, since many of the other small species of cat appear to be suitable for domestication (Cameron-Beaumont *et al.*, 2002).

The earliest solid archaeological evidence for domestic cats comes from the Egyptian New Kingdom, in about 1550 BC. By 850 BC the cult of the cat-goddess Bastet had given cats religious significance, and they were deliberately bred for sacrificial purposes. There were also periods during which cats were prohibited to be exported from Egypt (Serpell, 2000), though whether this arose from a recognition of their value as pest controllers, their sacred status or the former in the guise of the latter is unclear. Certainly cats did not become widespread in Europe until the 4th century AD.

The association between cats and paganism did not help their popularity in many parts of Western Europe from the time of the Reformation onwards, and some ritualistic persecution of cats persisted into the 20th century. Nevertheless, from the 19th century onwards the cat increased in popularity, and by the end of the 20th its numbers had surpassed those of the dog in many Western countries.

The majority of pet cats are of no particular breed, and are referred to variously as non-pedigree cats, mongrel cats, cross-bred cats and 'moggies'. Apart from the colour and length of their coats, which are likely to have been subjected to some deliberate selection by man (Todd, 1977), they vary little in size compared with most other domesticated species and, apart from ease of socialization to man, retain many of the characteristics of wildcats. Pedigree cats still make up a minority of the pet cat population. Two groups of breeds, the Siamese and Persian types, have been reproductively isolated from moggies for several centuries at least: the DNA of both of these indicates that they were derived from ordinary cats by mutation, rather than by cross-breeding with other subspecies or species of *Felis* (Driscoll *et al.*, 2007). More recently, breeds have been developed based on specific mutations, such as the Munchkin (dwarfism) the hairless Sphynx and the behaviourally unreactive Ragdoll. Others have been derived by hybridizing domestic cats with other felid species: examples include the Chausie (a hybrid with the jungle cat, *F. chaus*) and the Bengal, which derives its spotted coat and characteristic boldness from the Asian leopard cat, *Prionailurus bengalensis*.

15.2 Effects of Domestication

Excluding pedigree animals, the majority of cats are, in contrast with most other domesticated species, the product of unregulated matings between individuals that have selected one another. Population control is achieved largely by neutering, or by confining entire animals indoors; the latter is becoming more widespread in urban areas of the USA and parts of continental Europe, but is still uncommon in the UK and in rural areas. Hunting is a second wild-type behaviour retained by many cats; although few are very efficient predators, cats are effective enough as killers of birds and small mammals to have a local impact on their population dynamics (see below).

It is only recently that cats have been valued more for their companionship than for their pest-exterminating value but, until recently, it would have proved difficult to breed cats that showed little inclination to hunt. This stems from the cat's peculiar nutritional requirements (Legrand-Defrétin, 1994), which it shares with the whole of the Felidae. These include: (i) a restricted ability to shut down protein catabolism, resulting in a requirement for 18% protein in the diet of growing kittens; (ii) high dietary levels of sulphur-containing amino acids such as cysteine, methionine and taurine; (iii) an inability to synthesize prostaglandins from small precursor molecules; and (iv) a high requirement for niacin and thiamine. These present no problems when the cat's diet consists largely of raw carcases, but prevent cats from switching to a predominantly plant-based diet when prey is scarce, something that dogs, for example, are able to do. Until modern, nutritionally balanced cat foods became available, hunting was probably the only way that cats could guarantee themselves enough of these and other key nutrients to reproduce successfully, and has therefore persisted (Bradshaw *et al.*, 1999). Probably linked to this is the tendency of most outdoor cats to try to maintain an exclusive home range, an essential source of food for their recent ancestors.

15.3 Hunting Behaviour and Object Play

All cats, with the exception of some pedigrees, therefore have the potential to become effective predators, and this has not endeared them to people who are concerned about protecting wildlife. Their ecological impact is probably most serious when they have been introduced on to islands where there are no other mammalian predators (e.g. Bonnaud *et al.*, 2007), but the effect they have on urban bird populations, for example, is still debatable (Sims *et al.*, 2008). Even well-fed pet cats still go on hunting expeditions, thereby disrupting the behaviour of their prey species even if they are not successful at killing. Because cats do not hunt cooperatively, and are therefore restricted to small prey, if they are relying on prey for food they need to hunt before they become hungry, and hence feeding, hunger and predation are only loosely connected in the brain (Adamec, 1976).

Hunting is therefore, to a large extent, a 'free-running' behaviour, and the main influence of hunger appears to be in the size of prey that is approached, and the 'seriousness' with which it is attacked (Biben, 1979). Large or dangerous prey is only taken when the cat is especially hungry; when the cat is not hungry it may attack the prey but not kill it immediately. This strategy is presumed to have evolved from a trade-off between the immediate benefits of the food obtained from a kill and the risk of the cat becoming injured in the process. When these are balanced, the cat appears to 'play' with the prey item, although this is an anthropomorphic description for a behaviour more correctly labelled as 'motivated by conflict'. Conversely, the factors that induce adult cats to 'play' with toys suggest that, in motivational terms, this is predatory behaviour: the most effective toys resemble typical prey in size, surface texture and the way they move, and the intensity of the play and the size of toy preferred both increase with hunger (Hall, 1998).

15.4 Social Behaviour and Communication

The main difference between the wild *F. silvestris* and the domestic *F. catus* is in social behaviour, both towards other species – covered later on in this chapter – and towards members of its own species. *F. silvestris* is solitary and territorial, apparently even when its food is sufficiently clumped to permit the formation of a colony; the largest groups are therefore females with dependent offspring, which persist for only a few months. *F. catus* is capable of living at much higher density than its ancestor (Liberg *et al.*, 2000); this was presumably selected for during domestication, when granaries became large enough to attract more pests than one or two cats could control. This selection must have been complete before the Egyptians were able to breed large numbers of cats in their temples.

Our knowledge of the social behaviour of cats is largely derived from studies of free-ranging and feral groups formed around predictable and plentiful supplies of food, such as farmyards and fishing villages (Kerby and Macdonald, 1988). Here, cooperative behaviour is observed between females, usually close relatives; the most likely explanation for the establishment of these groups is an enhanced tolerance for female offspring, possibly selected for during domestication. Males rarely show any cooperative behaviour and are often nomadic, but they may remain attached to a colony, avoiding conflict with older individuals until they become strong and experienced enough to compete for mating opportunities.

The collaboration between females includes cooperative breeding, in which nests are shared, kittens are nursed and fed indiscriminately (see Fig. 15.1), and there is cooperative defence of the core area against unfamiliar cats of both sexes. Unfamiliar mature males may pose a risk of infanticide, although the extent to which this is an important cause of mortality in this species is unclear (Bonanni *et al.*, 2007). Like lions, but unlike wolves, all females within the group breed more or less simultaneously, without any reproductive suppression by 'dominant' individuals. Cooperation may be maintained by two tactile behaviour patterns, allorubbing (Kerby and

J. Bradshaw

Fig. 15.1. Kittens from three mothers suckling from one.

Macdonald, 1988; Fig. 15.2) and allogrooming, although neither of these has been studied as extensively as they have been in primates.

The signalling repertoire of cats largely reflects their sensory abilities (highly developed olfaction and hearing, vision insensitive to colour but tuned to movement detection) and their origins as solitary crepuscular animals (Bradshaw and Cameron-Beaumont, 2000). Scent marks, efficient for durable marking of territorial boundaries or locations of particular significance, are produced in: (i) urine, especially where this is spray-marked (see Fig. 15.3); (ii) faeces, if left uncovered; and (iii) in cheek gland secretions deposited at head height on projecting objects such as twigs (see Fig. 15.4). Reproductively active male cats ('toms') scent-mark particularly frequently and pungently, their urine containing high levels of sulphur-containing compounds that may be an indicator of quality. Social odours are perceived through both olfactory

Fig. 15.2. Feral cats allorubbing.

Fig. 15.3. Female cat spray-marking.

receptors and those in the vomeronasal organ, which cats bring into use through the 'flehmen' posture in which the mouth is opened slightly and the top lip is retracted.

While humans have difficulty appreciating its complexity and richness, cats undoubtedly rely a great deal on odours to inform themselves about their environment, both social and physical. By contrast, feral and free-ranging cats are generally silent, restricting their use of readily audible vocalizations to sexual and aggressive encounters and to mother–kitten interactions. Pet cats, especially Siamese and related breeds, are much more vocal. In particular, they use the 'miaow' much more frequently; this is probably a learned response, rewarded by humans through feeding, opening doors and so on. Some cats appear to have a repertoire of different miaows, each of which is reserved for a different context (Nicastro and Owren, 2003). Purring, the other characteristic vocalization of the pet cat, is a care-eliciting signal; there is no scientific evidence that it signifies the emotion of 'contentment' (Bradshaw, 1992).

As predicted from its ancestral solitary lifestyle, the domestic cat's repertoire of visual signals is restricted compared with that of the dog. The tail held vertically is a

Fig. 15.4. Cat scent-marking with its cheek glands.

J. Bradshaw

precursor to allorubbing and presumably indicates affiliative, rather than agonistic, intent. On the head, the pinnae (external ears) are the most expressive structures, pushed forward and erected to indicate confidence, and backwards to indicate intention to withdraw. Flattening of the ears indicates that the cat is expecting to engage in fighting. Pulled back and flattened are therefore a defensive posture, often accompanied by hissing. In agonistic encounters, cats will stand at full height, turn side-on and raise the fur on the back, to maximize their apparent size. Lashing of the tail, often accompanied by growling, is also a component of aggression. Individuals that are less confident may hunch their heads into their shoulders and flatten their bodies into the ground. Once two cats have begun to threaten one another, they find it difficult to de-escalate because they lack signals that indicate the acceptance of defeat: the loser will usually have to creep away very slowly, avoiding any rapid movement that might trigger a chase on the part of the more confident individual.

Cats within the same social group often rub on one another (allorubbing, Fig. 15.2), and the same behaviour pattern is frequently directed towards owners, or displaced on to objects near to people. As well as being a tactile signal, rubbing inevitably results in the exchange of some scent, although whether this is socially significant is not yet understood. The exchange of rubbing can be markedly asymmetric within a given pair of cats, leading to the suggestion that it may signify subordinate status between individuals that know one another well, though it is not a submissive signal in the sense that it is never performed in aggressive contexts. Mutual (allo)grooming is, however, sometimes followed by mild aggression (van den Bos, 1998), and may therefore be an attempt to assert dominance, the aggression resulting if this is disputed by the recipient of the grooming.

15.5   Courtship and Reproductive Behaviour

Female cats can be affiliative towards males at any time of year (Kerby and Macdonald, 1988), but reject all sexual advances except during the breeding season(s), typically, in the northern hemisphere, in late winter and midsummer. In pro-oestrus, females first increase their home ranges, and then begin to increase their rate of rubbing on objects. At this stage they presumably also emit odours that are attractive to males, which, in turn, show an abrupt increase in the size of the area they patrol, to encompass the female's range. As a result, several males may congregate around a female, but at this point she usually refuses to mate with any of them. Surprisingly, competition between males in such aggregations is usually low-key or non-existent (Liberg et al., 2000), although on occasion they will be dominated by a single male.

The female performs displays of rolling, purring, stretching and rhythmic protraction and retraction of the claws; after a day or two she comes into oestrus, and indicates her sexual receptivity by adopting the lordosis posture in between bouts of rolling. Typically she will be polyandrous, i.e. mate with several of the males that have aggregated around her, including the 'dominant' individual if there is one, but not exclusively. The attractiveness of the female for several days before she is sexually receptive suggests the function of assessing male quality, but this is not supported by the rather indiscriminate mating that follows. It has been suggested that this and other paradoxes in cat behaviour may be explained by the relatively short time, in

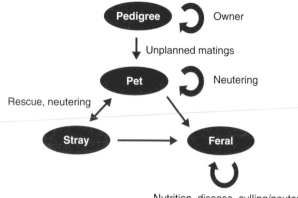

Fig. 15.5. Flow of genes between cat populations (arrows), with the main factors controlling reproductive success alongside each.

evolutionary terms, that they have lived at high density, such that the mating system is not yet optimized (Liberg *et al.*, 2000).

Apart from such details, it is apparent that most cats have retained wild-type sexual behaviour, enabling them to cross-breed with fully wild *F. silvestris* (Daniels *et al.*, 1998), and with feral cats. There may therefore be considerable gene flow between feral and domestic populations (see Fig. 15.5), with different selection pressures operating on each. As neutering of the pet population becomes more prevalent, so a higher proportion of genes from the largely un-neutered ferals will enter the pet population; since the adaptations required for survival as a feral may not be identical to those required to be a successful pet, this trend, if continued, may have unintended consequences for the pet population (Bradshaw *et al.*, 1999).

15.6 Dominance

Recently, it has been proposed that group-living cats exhibit a 'dominance hierarchy', based on ritualized signals and including the possibility of coalitions between weaker individuals to displace dominants from valued resources, such as occurs in primate groups (Crowell-Davis *et al.*, 2004). There is little doubt that, within most multi-cat households or feral groups, there is a loose 'peck order' in which some individuals have priority of access to food. However, there are differences of opinion as to which of the available signals indicate submission – for example, allorubbing (Kerby and Macdonald, 1988) or defensive behaviour (Natoli *et al.*, 2001). Different 'hierarchies' emerge from the same animals depending upon which behaviour patterns are used to construct them. If a hierarchy does actually exist in the minds of the cats, it appears (see above) to play little role in sexual or reproductive behaviour. Moreover, since the cat is derived from a solitary ancestor, the advanced social cognition needed to maintain sophisticated hierarchical systems would have to have been entirely selected for by domestication and, additionally, Liberg *et al.* (2000) consider that cats have yet to adapt fully to high-density living.

The peck orders that are observed may be more simply explained at the level of the dyad (pairwise relationship), using Parker and Rubenstein's (1974) resource-holding potential (RHP) model. In this conceptual framework, each animal assesses the other's apparent fighting ability in relation to its own and estimates the risks of escalating an encounter, counterbalanced by the benefit of winning. Thus the motivation to gain the resource under dispute plays an important role in determining the outcome of an encounter, explaining why a well-fed 'despot' will often allow a hungrier cat to feed. This approach can be applied equally to first encounters as to an established relationship, and does not require any complex cognitive abilities, such as the capacity to comprehend and remember third-party relationships. It is also widely applied to species which base their sociality on a territorial system (e.g. Barlow *et al.*, 1986), which domestic cats appear to retain or attempt to retain even when they are living at high densities, and may therefore be more explanatory for cats than the dominance concept.

15.7 Ontogeny of Behaviour

The behavioural development of kittens is conventionally divided into four phases: (i) the prenatal period; (ii) the neonatal period, which lasts until the end of the second week post-gestation, during which the kitten is relatively helpless and not yet sensorially competent; (iii) the socialization period from the third to eighth weeks, when the kitten actively engages in, and learns about, social relationships; and (iv) the juvenile period, when it becomes increasingly independent, ending with the transition to adulthood at the onset of sexual maturity.

Kittens are born at an early stage of development, blind, virtually deaf and incapable of thermoregulation, although the senses of touch, taste and smell are quite well developed. Accordingly, they use mainly odour and tactile cues on the mother's abdomen reflexively to locate a nipple and begin to suckle. Suckling is often accompanied by purring, presumably a signal to the mother that inhibits her from moving off. Initially, kittens cannot walk and pull themselves along with rowing motions of the forelimbs, orientating themselves largely by odour. The hindlimbs mature during the third and fourth weeks of life, enabling the kitten to start walking and then running; full coordination is achieved in the seventh week. Voluntary urination and defecation are not possible until about the fourth week, prior to which the mother disposes of all the kittens' wastes during grooming.

The kitten's hearing develops gradually, as the ear canals open up and the external ears (pinnae) unfold and become upright during the second week of life. Vocalizations made by other cats elicit more interest than other sounds, and become integrated with locomotory responses during the fourth week; maternal calls are approached, agonistic vocalizations are avoided.

The timing of the opening of the eyes is very variable, occurring almost any time in the first fortnight of life, depending on genetic factors, the sex of the kitten and possibly the level of light in the environment. The optic fluid, initially cloudy, clears gradually, allowing improved image formation. The visual cortex develops in parallel with the eyes; initially, many of the cells respond to both eyes, but they rapidly become specialized to one or the other, although retaining connections to allow binocular vision. This process is not spontaneous, but requires visual experience, as does the development of specialized outline- and movement-detection neurons that are thought to be an essential part of the cat's prey-detection systems. As for binocular vision, the initial cells are rather non-specific, but these guide their own replacement by more specialized cells, specific to vertical or horizontal edges, circular areas of contrast and movement in various dimensions and directions. If the initial non-specific neurons are not stimulated they degenerate spontaneously, such that the visual system cannot develop from scratch after the kitten is about 3 months old.

From the fourth week of life onwards, kittens spend much of their waking time playing. Initially, much of this is directed at littermates, but this may simply be because these are the most interesting objects in the den, apart from the mother. By 7 weeks of age, object and social play have become distinct, not only in terms of the target they are directed at, but also in the behaviour patterns employed. Play with objects involves many patterns that are subsequently incorporated into hunting behaviour, while signals such as the open mouth, sidestep and face-off (West, 1974) are used mainly or exclusively in social play. Subsequent development of play is complex and varies from one litter and individual to another (Bateson, 2000), as might be expected for a 'luxury' behaviour, opportunities for which will be affected by the quality of the environment in which the cat finds itself.

The behaviour of the mother is crucial to the success of the litter, initially in selecting a suitable denning site, since kittens have to be kept warm and dry for the first few weeks of life if they are to survive. Mothers do not build a nest as such, but they usually choose a well-protected location in which to give birth, and may temporarily become aggressive towards other cats or household animals, presumably to reduce the risk of predation. After a few weeks the kittens may be moved, one at a time, to a new den, presumably for reasons of hygiene and/or to reduce ectoparasite transmission. The mother takes advantage of the 'scruff' reflex: grasping the kittens by the loose skin behind their necks causes them to curl up but inhibits all other behaviour. The same reflex is used by male cats to subdue females at intromission, and as a benign method by humans for immobilizing cats (Pozza et al., 2008).

Throughout the period up to weaning, the mother's behaviour provides essential elements for the kittens' development; without her involvement, they are likely to develop a range of physical, emotional and behavioural abnormalities (Bateson, 2000). For example, in the wild she will begin to bring back half-killed prey to the nest, once the kittens are about 4 weeks old: this is an important component in their acquisition of hunting skills. Soon after this she will begin to withdraw from the kittens for increasing periods of time, stimulating weaning; as a result, there is often a temporary check in the weight gain of the kittens. Lactation is metabolically costly, probably critically so for a mother that is subsisting by hunting, and so it pays the mother to wean her kittens as early as possible, so that she is not weakened herself. If she has been underweight during gestation, she may become aggressive towards her kittens at this point. The optimum strategy for each kitten, on the other hand, is to prolong lactation and thereby not expose itself to the risks of leaving the den until older and stronger. However, this 'maternal–infant conflict' is apparent only in large litters, and lasts only for a few days, after which the kittens continue to suckle for several more weeks, albeit at a lower rate than before.

15.8 Socialization

The term 'socialization' is used rather loosely to describe a diverse set of developmental processes whereby young animals learn how to both identify and interact with members of their own species, and also how to interact with members of other species, almost always man. Some, but probably not all, of these processes are analogous to 'imprinting' and take place most readily during 'sensitive periods' (Bateson, 2000; Turner, 2000). In the domestic cat, the sensitive period for socialization to humans begins at about 2 weeks of age, as soon as the relevant sensory systems become competent; domestic kittens which have not been handled by people until they are 8 or more weeks old are subsequently much less sociable to people (Turner, 2000), indicating that this sensitive period may spontaneously come to an end at about that time, probably because the kitten then becomes fearful of the unfamiliar. Kittens handled from 2 weeks of age are friendlier towards people at 5 weeks old than kittens not handled until that point, but if both are handled from 5 weeks onwards they become behaviourally very similar by 6 months of age (Lowe and Bradshaw, 2002); very early handling is therefore facilitating, but not essential, for socialization.

Quantity and quality of handling also have long-term effects (Casey and Bradshaw, 2008). For kittens which do not develop such a reaction, learning about how to interact with people appears to continue for several more weeks (McCune, 1995; Lowe and Bradshaw, 2001, 2002), into the juvenile period. The sensitive period appears to be terminated by some internal event, in that there is no evidence that it continues if positive social interaction with people does not occur by week 8; it could be terminated by a developmental clock, or by the completion of a sensitive period for intraspecific socialization.

It is generally assumed that intraspecific socialization, i.e. to other cats, occurs simultaneously with socialization to people (Bateson, 2000). Since the ancestral species is solitary, intraspecific socialization in the domestic cat presumably evolved to establish social bonds with the mother and litter-mates, and possibly to direct subsequent sexual preferences. Anecdotally, cats vary considerably and persistently in their tolerance of conspecifics, Mertens and Schär (1988) arguing that intraspecific sociability is a stable characteristic of individuals. Possible underlying genetic influences may include the 'boldness' trait that affects socialization to humans (McCune, 1995); ontogeny is a more likely explanation for the greater tolerance of litter-mates for close confinement with each other compared with cohabiting, but unrelated, cats (Bradshaw and Hall, 1999).

The sensitive period for bonding with the mother (filial imprinting) appears to peak in the fourth and fifth weeks (Schneirla and Rosenblatt, 1961), which is compatible with simultaneous intra- and interspecific socialization. However, Guyot *et al.* (1980) suggested that interaction with litter-mates is more important in the development of cooperative social interactions than interaction with the mother. Social play usually declines after 16 weeks of age, suggesting that a sensitive period for the formation of relationships with individuals other than the mother may be coming to an end at about this age. Subsequent tolerance of other cats could be based upon either: (i) the memorization of the characteristics of individuals, as presumably occurs earlier with the mother; or (ii) a general tendency to be tolerant of all unfamiliar individuals if a wide variety of cats has been encountered during the sensitive period. Functionally, both of these seem plausible: based on the pattern of cooperation between related females, individuals encountered in the first few months of life are likely to be kin and therefore bring mutual benefit to subsequent cooperation; kittens meeting non-siblings during the same period are likely to have been born in an environment where food is both plentiful and reliable enough to favour group-living over territoriality in subsequent years also.

As in the dog, there is no evidence to indicate that 'socialization' to humans interferes with the development of intraspecific social behaviour; kittens raised with both conspecifics and humans subsequently direct their sexual behaviour only towards conspecifics.

The quality and quantity of socialization received by an individual is a major contributor to its 'personality', a set of behavioural biases which are repeatable over time and across situations, including the shy/bold continuum that is evident in animals from many taxa (Réale et al., 2007). Genetic influences are also detectable in cats (McCune, 1995); experimentally these are easiest to detect as paternal effects, because the male plays no part in rearing kittens, but it is assumed that mothers have an equivalent genetic input, in addition to direct and indirect influences on what their kittens learn (McCune et al., 1995). Cat 'personalities' appear to become stable after about 2 years of age (Lowe and Bradshaw, 2002), prior to which they remain open to environmental influences and learning opportunities.

15.9 The Ethology of Cat–Human Relationships

Cat–human relationships, although they have not received the same level of attention as interactions between humans and dogs, have been studied from a wide range of perspectives, only some of which have an ethological component. Those that fall outside the scope of this book include the influences of cat ownership on human health (McNicholas et al., 2005), attachment and social support (Stammbach and Turner, 1999) and the psychology of stray cat feeding (Toukhsati et al., 2007). Ethological (observational) methods have occasionally been used to study cat–owner relationships (summarized in Turner, 2000), but the interpretation has usually been based on psychological constructs.

Cat behaviour 'problems', as presented clinically, can usefully be divided into those which are species-typical behaviour performed in an inappropriate context and those which involve some kind of pathology, although there is overlap between these (Casey et al., in press). The most common behaviour presented clinically is inappropriate elimination ('house soiling': Bamberger and Houpt, 2006), and this is often 'normal' scent-marking behaviour performed in an unwelcome place. Other cat behaviour 'problems' that derive from normal ethology include territorial aggression, aggression towards cats within an artificially constituted social group, defensive aggression towards owners, withdrawal and hiding behaviour, scratching household objects and unwanted predatory behaviour (see Casey et al., in press, for further details).

References

Adamec, R.E. (1976) The interaction of hunger and preying in the domestic cat (*Felis catus*): an adaptive hierarchy? *Behavioural Biology* 18, 263–272.
Bamberger, M. and Houpt, K.A. (2006) Signalment factors, comorbidity, and trends in behavior diagnoses in cats: 736 cases (1991–2001). *Journal of the Veterinary Medical Association* 229, 1602–1606.

Barlow, G.W., Rogers, W. and Fraley, N. (1986) Do Midas cichlids win through prowess or daring? *Behavioral Ecology and Sociobiology* 19, 1–8.

Bateson, P. (2000) Behavioural development in the cat. In: Turner, D.C. and Bateson, P. (eds) *The Domestic Cat: the Biology of its Behaviour*, 2nd edn. Cambridge University Press, Cambridge, UK, pp. 9–22.

Biben, M. (1979) Predation and predatory play behaviour of domestic cats. *Animal Behaviour* 27, 81–94.

Bonanni, R., Cafazzo, S., Fantini, C., Pontier, D. and Natoli, E. (2007) Feeding-order in an urban feral domestic cat colony: relationship to dominance rank, sex and age. *Animal Behaviour* 74, 1369–1379.

Bonnaud, E., Bourgeois, K., Vidal, E., Kayser, Y., Tranchant, Y. and Legrand, J. (2007) Feeding ecology of a feral cat population on a small Mediterranean island. *Journal of Mammalogy* 88, 1074–1081.

Bradshaw, J.W.S. (1992) *The Behaviour of the Domestic Cat*. CAB International, Wallingford, UK.

Bradshaw, J.W.S. and Cameron-Beaumont, C. (2000) The signalling repertoire of the domestic cat and its undomesticated relatives. In: Turner, D.C. and Bateson, P. (eds) *The Domestic Cat: the Biology of its Behaviour,* 2nd edn. Cambridge University Press, Cambridge, UK, pp. 67–93.

Bradshaw, J.W.S. and Hall, S.L. (1999) Affiliative behaviour of related and unrelated pairs of cats in catteries: a preliminary report. *Applied Animal Behaviour Science* 63, 251–255.

Bradshaw, J.W.S., Horsfield, G.F., Allen, J.A. and Robinson, I.H. (1999) Feral cats: their role in the population dynamics of *Felis catus. Applied Animal Behaviour Science* 65, 273–283.

Cameron-Beaumont, C., Lowe, S.E. and Bradshaw, J.W.S. (2002) Evidence suggesting pre-adaptation to domestication throughout the small Felidae. *Biological Journal of the Linnean Society* 75, 361–366.

Casey, R.A. and Bradshaw, J.W.S. (2008) The effects of additional socialization for kittens in a rescue centre on their behaviour and suitability as a pet. *Applied Animal Behaviour Science* 114,196–205.

Casey, R.A., Blackwell, E.-J. and Bradshaw, J.W.S. (in press) *Principles of Companion Animal Behaviour Therapy.* Wiley-Blackwell, Oxford, UK.

Crowell-Davis, S.L., Curtis, T.M. and Knowles, R.J. (2004) Social organisation in the cat: a modern understanding. *Journal of Feline Medicine and Surgery* 6, 19–28.

Daniels, M.J., Balharry, D., Hirst, D., Kitchener, A.C. and Aspinall, R.J. (1998) Morphological and pelage characteristics of wild living cats in Scotland: implications for defining the 'wildcat'. *Journal of Zoology* 244, 231–247.

Diamond, J. (1997) *Guns, Germs and Steel: a Short History of Everybody for the Last 13,000 Years*. Jonathan Cape Ltd., London, UK.

Driscoll, C.A., Menotti-Raymond, M., Roca, A.L., Hupe, K., Johnson, W.E., Geffen, E., Harley, E.H., Delibes, M., Pontier, D., Kitchener, A.C., Yamaguchi, N., O'Brien, S.J. and Macdonald, D.W. (2007) The Near Eastern origin of cat domestication. *Science* 317, 519–523.

Guyot, G.W., Cross, H.A. and Bennett, T.L. (1980) Early social isolation of the domestic cat: responses to separation from social and nonsocial rearing stimuli. *Developmental Psychobiology* 13, 309–315.

Hall, S.L. (1998) Object play by adult animals. In: Bekoff, M. and Byers, J.A. (eds) *Animal Play: Evolutionary, Comparative and Ecological Perspectives*. Cambridge University Press, Cambridge, UK, pp. 45–60.

Kerby, G. and Macdonald, D.W. (1988) Cat society and the consequences of colony size. In: Turner, D.C. and Bateson, P. (eds) *The Domestic Cat: the Biology of its Behaviour*. Cambridge University Press, Cambridge, UK, pp. 67–82.

Legrand-Defrétin, V. (1994) Differences between cats and dogs: a nutritional view. *Proceedings of the Nutrition Society* 53, 15–24.

Liberg, O., Sandell, M., Pontier, D. and Natoli, E. (2000) Density, spatial organisation and reproductive tactics. In: Turner, D.C. and Bateson, P. (eds) *The Domestic Cat: the Biology of its Behaviour*, 2nd edn. Cambridge University Press, Cambridge, UK, pp. 117–147.

Lowe, S.E. and Bradshaw, J.W.S. (2001) Ontogeny of individuality in the domestic cat in the home environment. *Animal Behaviour* 61, 231–237.

Lowe, S.E. and Bradshaw, J.W.S. (2002) Responses of pet cats to being held by an unfamiliar person, from weaning to three years of age. *Anthrozoös* 15, 69–79.

McCune, S. (1995) The impact of paternity and early socialisation on the development of cats' behaviour to people and novel objects. *Applied Animal Behaviour Science* 45, 109–124.

McCune, S., McPherson, J.A. and Bradshaw, J.W.S. (1995) Avoiding problems: the importance of socialisation. In: Robinson, I. (ed.) *The Waltham Book of Human–Animal Interaction: Benefits and Responsibilities of Pet Ownership.* Elsevier Science Ltd., Oxford, UK, pp. 71–86.

McNicholas, J., Gilbey, A., Rennie, A., Ahmedzai, S., Dono, J.-A. and Ormerod, E. (2005) Pet ownership and human health: a brief review of evidence and issues. *British Medical Journal* 331, 1252–1254.

Mertens, C. and Schär, R. (1988) Practical aspects of research on cats. In: Turner, D.C. and Bateson, P. (eds) *The Domestic Cat: the Biology of its Behaviour.* Cambridge University Press, Cambridge, UK, pp. 179–190.

Natoli, E., Baggio, A. and Pontier, D. (2001) Male and female agonistic and affiliative relationships in a social group of farm cats (*Felis catus* L.). *Behavioural Processes* 53, 137–143.

Nicastro, N. and Owren, M.J. (2003) Classification of domestic cat (*Felis catus*) vocalisations by naïve and experienced human listeners. *Journal of Comparative Psychology* 117, 44–52.

Parker, G.A. and Rubenstein, D.I. (1974) Role assessment, reserve strategy, and acquisition of information in asymmetrical animal conflicts. *Animal Behaviour* 29, 221–240.

Pozza, M.E., Stella, J.L., Chappuis-Gagnon, A.-C., Wagner, S.O. and Buffington, C.A.T. (2008) Pinch-induced behavioral inhibition ('clipnosis') in domestic cats. *Journal of Feline Medicine and Surgery* 10, 82–87.

Randi, E. and Ragni, B. (1991) Genetic variability and biochemical systematics of domestic and wildcat populations (*Felis silvestris*: Felidae). *Journal of Mammalogy* 72, 79–88.

Réale, D., Reader, S.M., Sol, D., McDougall, P.T. and Dingemanse, N.J. (2007) Integrating animal temperament within ecology and evolution. *Biological Reviews* 82, 291–318.

Schneirla, T.C. and Rosenblatt, J.S. (1961) Behavioral organisation and genesis of the social bond in insects and mammals. *Journal of Orthopsychiatry* 31, 223–253.

Serpell, J.A. (2000) Domestication and history of the cat. In: Turner, D.C. and Bateson, P. (eds) *The Domestic Cat: the Biology of its Behaviour*, 2nd edn. Cambridge University Press, Cambridge, UK, pp. 179–192.

Sims, V., Evans, K.L., Newson, S.E., Tratalos, J.A. and Gaston, K.J. (2008) Avian assemblage structure and domestic cat densities in urban environments. *Diversity and Distributions* 14, 387–399.

Stammbach, K.B. and Turner, D.C. (1999) Understanding the human–cat relationship: human social support or attachment. *Anthrozoös* 12, 162–168.

Todd, N.B. (1977) Cats and commerce. *Scientific American* 237 (November), 100–107.

Toukhsati, S.R., Bennett, P.C. and Coleman, G.J. (2007) Behaviors and attitudes towards semi-owned cats. *Anthrozoös* 20, 131–142.

Turner, D.C. (2000) The human–cat relationship. In: Turner, D.C. and Bateson, P. (eds) *The Domestic Cat: the Biology of its Behaviour*, 2nd edn. Cambridge University Press, Cambridge, UK, pp. 193–206.

van den Bos, R. (1998) The function of allogrooming in domestic cats (*Felis silvestris catus*); a study in a group of cats living in confinement. *Journal of Ethology* 16, 1–13.

West, M.J. (1974) Social play in the domestic cat. *American Zoologist* 14, 427–436.

The Behaviour of Laboratory Mice and Rats

H. WÜRBEL, C. BURN AND N. LATHAM

16.1 General Biology

Mice and rats belong to the order Rodentia, which forms the largest and most diverse group of mammals, comprising approximately 1700 species, or 40% of all known mammalian species. Their common feature is a unique gnawing action, provided by massive masseter jaw muscles and sharp, ever-growing incisor teeth. Gnawing allows rodents to eat the toughest nuts and seeds and to gnaw through wood and roots in search of food and shelter. Most rodents are small-bodied, although the largest species, capybara (*Hydrochaerus hydrochaeris*), may get as large as 60 cm high and 130 cm long. Most rodents are frugivorous/herbivorous, but some are omnivorous (e.g. most of the genus *Rattus*) or insectivorous (e.g. the desert dormouse, *Selevinia betpakdalaensis*).

Rodents inhabit virtually every type of terrestrial habitat. Some species are arboreal (e.g. arboreal squirrels, New World porcupines); others dig extensive burrow systems (e.g. gerbils, mole rats, ground squirrels). Most rodents are nocturnal, depending on secluded shelters providing protection from predators (e.g. mice, rats, voles, hamsters, gerbils), while those living in less predated areas are diurnal and nest above ground (e.g. guinea pigs, chinchillas). However, individual species vary greatly in the range of habitats they occupy, reflecting different degrees of adaptability between them. Among the extreme generalists are the house mouse (*Mus musculus*) and the Norway rat (*Rattus norvegicus*). Their adaptability has facilitated adaptation to human habitation, allowing them to exploit rich food sources. In turn, it has predisposed them for use as laboratory animals. Today, mice and rats account for about 85% of all animals used in research worldwide. With an estimated 17–22 million vertebrates used in the USA and about 12 million in the EU, each year about 30 million rodents are used in animal experiments alone in these two parts of the world.

16.2 Origin and Domestication History

The mouse

The laboratory mouse is descended from one of the most widespread and successful mammal species, the house mouse (*Mus musculus*). House mice have had a close relationship with humans dating back many thousands of years. They originated on the steppes of Asia, and are thought to have spread around the world with human migration, now being found in almost every region of the planet. House mouse populations are typically defined in one of two ways based on their dependence on humans: *commensal* mice live in man-made environments, while *feral* mice live more independently of humans, in natural environments. As illustrated below, this difference has a number of effects on house mouse behaviour.

Mice have been used in research for over 300 years but the inbred laboratory mouse (as we know it today) first appeared a century ago, with the first inbred strain (DBA) being created by Clarence Cook Little at Harvard University in the USA, in order to study the inheritance of coat colour. Today, the laboratory mouse is the most widely used vertebrate species in biomedical research, with over 1000 genetically defined inbred strains (including the widely used C57BL/6 and Balb/c strains), and increasing numbers of genetically altered strains with either gene deletions (knockout) or insertions (transgenics). Numerous outbred strains are also available (including ICR CD-1), which are generally considered more genetically heterogeneous than their inbred counterparts. However, even these strains are derived from very small numbers of progenitor animals. Besides their adaptability, a short reproductive cycle, short life-span, small body size and low cost of maintenance were the main features predisposing mice for 'success' as laboratory animals.

The rat

Having originated in northern China, the Norway rat, like the house mouse, has spread to all continents, except Antarctica, and is now the dominant rat species in Europe and most of North America. Its name is actually a misnomer, going back to the English naturalist John Berkenhout, who gave the brown rat the taxonomic name *Rattus norvegicus*, believing that it had migrated to England on Norwegian ships in the early 1700s.

Laboratory rats were first established in 1895 from a population of albino rats at Clark University in Worcester, Massachusetts. Originally used for research on diet and physiology, rats quickly became used in all areas of biomedical research and they are the predominant animal model in experimental psychology.

Domestic rats are calmer and less likely to bite than their wild ancestors. They tolerate greater crowding, breed earlier and produce more offspring and, as in most domestic species, their brains are smaller.

Like laboratory mice, laboratory rats are available as outbred and (albeit less frequently than in mice) inbred strains. Most strains were derived from the Wistar rat. Other popular strains are Sprague Dawley, Fischer 344, Long-Evans and Lister hooded rats.

It was not until recently that knockout and transgenic rats became available, because the simple techniques that had worked in mice did not work in rats. Meanwhile, genetically modified rats are increasingly popular because it is widely believed that they model human physiology better than do mice.

16.3   Diurnal Rhythm, Sleep and Grooming Behaviour

The mouse

House mice are generally crepuscular or nocturnal, but their activity patterns may be altered by factors such as the timing of food availability, human activity in man-made environments and reproductive state, e.g. lactation (Latham and Mason, 2004). Little is known about the total or individual duration of activity bouts in free-living mice.

Observations of laboratory mice show that they can be continuously active for several hours at a time, but that they are active for less than 50% of each 24 h period (Baumgardner *et al.*, 1980) – although this is probably an underestimate of activity in free-living mice, whose environments are far more demanding.

Inactive periods are spent in sheltered nests which provide protection against predators and adverse weather. In feral environments nests are usually constructed in underground burrows, and may be lined with grass, hair or feathers. Mice are efficient tunnellers in soft earth, and burrows may consist of simple tunnels up to 1 m long, or complex tunnel systems with numerous chambers and exits. In commensal environments, mice build spherical or bowl-shaped nests consisting of a loose, outer paper- or rag-based structure and lined with finer shredded material (Randall, 1999). Mice typically nest either alone or with their breeding partner and family group, although under adverse conditions (e.g. extreme cold) even highly territorial individuals will nest with other mice, including subordinates that they would normally attack (Crowcroft, 1966).

The transition between inactivity and activity usually involves a period of grooming, starting with the head and progressing caudally. Grooming also occurs sporadically during activity bouts and, in captivity at least, it can occupy roughly 17% of the time budget (Baumgardner *et al.*, 1980). Self-grooming is important for hygiene and maintaining the coat for insulation, while allogrooming helps to maintain social relationships. Grooming can also occur particularly intensively after eating and may then, during allogrooming, assist in the transfer of information about foodstuffs (Crawley, 2000).

The rat

The circadian rhythm of Norway rats is most accurately described as being crepuscular, with most activity occurring just after dusk and before dawn. However, although this is certainly their preference, they can adjust their rhythm relatively flexibly depending on food availability, predation risk, weather and sexual activity, so it is not uncommon to see rats in daylight (Calhoun, 1963; Fig. 16.1). They do not hibernate but, in temperate climates, are more active in warmer months than during the winter.

Rats sleep within underground burrows in nests made of grass or other vegetation, or shredded paper if available (Calhoun, 1963). The nests are usually flat or cup-shaped pads within chambers of 15–40 cm in diameter, but are rarely as complex as mouse nests. Provided that fewer than five or six rats share a nest, it is seldom soiled. The burrow itself consists of tunnels with a mean diameter of 8–10 cm and an entrance, which can be sealed temporarily with vegetation or, more permanently, with soil. The straight sections of the tunnels have median lengths around 30 cm, beyond which they may bend, form cavities or split into two tunnels. Despite their apparently delicate forepaws, rats are effective burrowers, preferring loose, well-drained soil that does not crumble, but their burrows are shallower than those of many subterranean rodent species.

Most grooming takes place in the safety of the burrow. Rats groom daily, presumably removing dirt, reducing ectoparasite loads and distributing their own scents and conditioning oils evenly over the coat. Grooming is often associated with resting, but it can also be provoked by novelty, disturbance or frustration. Laboratory studies

Fig. 16.1. Rats are good climbers. Although mostly nocturnal, it is not unusual to see rats during daylight depending on food availability, predation risk, weather and sexual activity (image courtesy of Charlotte Burn).

show that grooming starts with the head and progresses caudally, but grooming bouts are shorter and rarely progress to the caudal regions when rats are more stressed and vigilant.

16.4 General Activity, Foraging and Feeding Behaviour

The mouse

House mice are creatures of habit, especially commensal mice, where territories or home ranges tend to be smaller than those of their feral counterparts (Latham and Mason, 2004). When out and about mice use visual landmarks to navigate around their environment and repeatedly follow familiar routes, leading to well-worn runways (Randall, 1999). Mice are also extremely agile, being excellent climbers, able to scale vertical brickwork or bark; they are proficient jumpers, able to leap over 30 cm vertically and 2.5 m downwards without harm, and they are capable swimmers (Randall, 1999). However, mice are thigmotactic (staying near walls), and will generally avoid areas that have been cleared of vegetation, exposed areas or areas regularly disturbed by human activity (Meehan, 1984).

Regular, predictable movement around the environment serves a number of functions. It allows the investigation of olfactory cues used in communication, which contain information about age, sexual status, relatedness and individual identity. These cues also play key roles in territorial, sexual and parental behaviour, as discussed below. Many of these cues are found in the urine (although important cues are also present in faeces and plantar gland secretions). Urinary odour cues may last for

H. Würbel *et al.*

up to 2 days, because the volatile signalling pheromones are bound to non-volatile major urinary proteins (MUPs), which act as a slow-release mechanism (Hurst *et al.*, 2001). Mice also use regular movements around their territories to monitor the presence of predators, which include owls and weasels in natural environments, and the domestic cat in urban areas.

Mice also incorporate foraging into their movements around their environment. Mice eat up to 20% of their body weight daily, consuming about 200 small meals per night from up to 30 food sites (Meehan, 1984). Mice are omnivores, but show preferences for foods high in fat and protein. The diet of commensal mice is often determined by the nature of their environment, e.g. grain in grain stores, while the diet of feral mice can be very varied, and includes cereals, grasses, roots and seeds (Berry, 1970). House mice also exhibit predatory behaviour, eating live insects and their larvae, and even seabird chicks (Latham and Mason, 2004). House mice acquire food preferences from their mothers and from other mice through the transfer of odour cues during allogrooming, and are generally cautious about trying novel foods unless they have been consumed by other individuals. Mice generally meet their water requirements from their food, if consuming diets with moisture contents of at least 15% – but they will drink free water if it is available, and require additional water if living in hot, arid environments or feeding on a dry or protein-rich diet (Meehan, 1984).

The rat

Rats are opportunistic omnivores, consuming diverse foods and employing varied methods of foraging: factors that have led to them becoming major pests. Wild rats choose to inhabit areas with available running water (their foremost priority), accessible food, exposed soil for digging a burrow, and cover. They avoid areas densely populated with humans, unless food is particularly plentiful there, such as through inadequate waste disposal. Rats patrol their territory regularly, travelling along runs that usually follow vertical features, such as walls – rats are thigmotactic, so they maintain physical contact with vertical surfaces whenever possible. The range size is usually about 30–45 m in diameter (Davis, 1953), but can be up to about 150 m.

Because of their opportunistic lifestyle, rats often forage even when satiated, eating little but learning the whereabouts of food for future reference. Rats locate food mainly through olfaction, which is several orders of magnitude more sensitive than our own. They can even locate the direction of an olfactory source without moving their heads by three orders of magnitude more quickly than humans can. Wild rats forage by digging in loose substrates such as leaf litter, by climbing trees and even by diving into streams to collect molluscs. They are central-place foragers, meaning that once food is found they usually carry it in their mouths to the burrow or a nearby refuge. There they consume the food, holding it in their forepaws. Their teeth allow them to chew through tough materials, such as nuts and snail shells, and even lead piping (Barnett, 2005).

Their diet can include fruit, seeds, nuts, molluscs, eggs, small vertebrates and any food waste left by humans. In fact, they have similar food preferences to humans, finding sweet, calorific and high-protein foods particularly attractive, consuming salty or acidic foods in moderation and avoiding bitter foods (Burn, 2008). They can even acquire learned preferences for chilli-flavoured foods, despite initially avoiding them. However, some species differences exist; for example, rats taste the harmless compound denatonium benzoate as being less bitter than humans and many other animals do. Therefore, this can be added to poisoned baits, discouraging non-target species from consuming them, while rats remain relatively undeterred.

Rats cannot vomit, but laboratory studies show that consumption of toxins makes them more likely to consume non-nutritional substances, such as clay, possibly serving to 'neutralize' the toxins. Wild rats avoid consuming lethal quantities of poisons, primarily through their neophobic behaviour. That is, they avoid eating unfamiliar foods if familiar ones are available, and will initially eat only very small quantities of unfamiliar foods, if any. They then wait several hours before eating more of that food and, if they suffer any ill effects from it, they subsequently avoid eating it. This learned avoidance is a very powerful form of one-trial learning, and rats may continue to avoid the food for months if not years. A lactating mother rat will even avoid a novel food if her pups become sick from the milk produced after she ate it. Calhoun (1963) fed rats in a semi-natural enclosure on standard chow, but occasionally supplied mixed 'garbage', which they initially approached with the stretched–attend posture, consuming little. However, once accustomed to the garbage, they concentrated on eating and caching it, only returning to eating the chow once it was finished.

As well as learning about foods from their own experiences, rats can learn about it socially (Burn, 2008). Rats prefer novel foods after smelling them on conspecifics' breath, an effect mediated through carbon disulphide (present in rats' breath) (Galef *et al.*, 1988), so new food preferences can spread rapidly through rat colonies. Rats also avoid foods that they themselves have eaten if they encounter the odour of a sick conspecific soon after eating, but this does not prevent them from eating the food that might have made the conspecific sick; in fact carbon disulphide on the sick rats' breath still induces a preference for that food.

16.5 Social Organization and Social Behaviour

The mouse

The social structure and territorial behaviour of house mouse populations vary according to the animals' environment. In feral conditions both male and female mice may hold loosely defended overlapping territories (or home ranges). Home range sizes often vary with the density of available food, with reported densities ranging from one mouse per 3–33 m² in fields of grain – populations in some agricultural areas increase fivefold in the weeks immediately following crop sowing (Latham and Mason, 2004). However, in some areas – for example, the wheatlands of Australia – mice may have home ranges up to an incredible 80,000 m². In contrast, in commensal environments territories are typically held only by males (although a male's breeding mate(s) and juvenile offspring may assist in territorial behaviour) (Crowcroft, 1966). Territories are also much smaller and more strictly defended.

Territorial boundaries often occur at physically and/or visually striking features of the environment. Territory ownership is signalled, particularly by dominant males, via regularly refreshed urine marks along the borders of their territory (Hurst *et al.*, 2001). These marks are also laid down throughout the rest of the territory, especially on conspicuous objects and favoured nesting and feeding sites. Dominant males may even venture into neighbouring territories to over-mark the urine marks of their competitors, as fresh urine marks signal competitive ability. Dominant males will not tolerate marks from other adult males (i.e. potential competitors), but will generally tolerate marks from juveniles and females. This is thought to be because juvenile urine lacks the aggression-eliciting properties of adult male urine, and female urine contains an aggression-inhibiting factor.

When male mice meet they use plantar, preputial and salivary gland scents to identify each other and to assess each other's 'maleness' and sexual state. At high population densities (usually in commensal populations) territorial males will chase intruders out of their territories, biting at their rump and tail. The intruder may indicate his subordinancy by standing, raising his forepaws and exposing his belly (Crawley, 2000). However, attacks can be intense and sustained and, if unable to escape, the intruder is likely to face severe injury and even death. In feral populations, territory holders are generally more tolerant of known subordinates, and will vigorously defend only favoured nesting and feeding sites.

The rat

Rats are highly social. Colony sizes can vary widely, from one male and female with their offspring to colonies of 200 or more individuals, depending on resource availability and physical boundaries. An 18-month study (Traweger *et al.*, 2006) of wild rats in Salzburg, Austria, found that densities ranged from ten rats per 1 km along the banks of standing waters up to 113 rats per 1 km of riverbank, but in many areas of the city none were trapped at all.

Rat colonies are stable, with low migration rates, provided the environment allows (Davis, 1953), so overt aggression towards established colony members is relatively rare. Most aggression occurs between young adult male rats as they establish their place in the dominance hierarchy, and towards intruders. Older (usually larger) rats are dominant over younger ones. When population densities are low, rats live in territorial groups of one dominant male, up to about six females, and their offspring. When the density is higher, the dominant male can no longer maintain a monopoly on the females and many subordinate males can cohabit within the colony. Subordinates of both sexes usually face restricted access to food and other resources, have significantly more wounds than dominants and have higher mortality rates (Davis, 1953).

Rats recognize familiar individuals largely through their individual olfactory characteristics, which can be determined genetically or acquired from the environment (Burn, 2008). Much olfactory information is mediated through urine, but rats have many other specialized scent glands. They can thus gain information about each other's gender, reproductive state, genetic relatedness, dominance, health status and individual identity through sniffing each other or their urine. Dominant males in particular use urine marking to delineate their territory and to maintain their dominant status. Rats communicate with each other using scents, including alarm or reward odours that they deposit in harmful or beneficial environments, respectively. They also communicate using vocalizations – both audible to humans and ultrasonic – and using postures.

Aggression is strongly mediated through scent (Burn, 2008), with males that smell unfamiliar (i.e. usually intruders) being attacked; sometimes, unfamiliar females

---

are also attacked. Male offensive attack includes lateral attack, bites targeted at the opponent's rump, chasing and pinning the opponent down (Blanchard *et al.*, 2001). Defensive attack involves bites directed at the opponent's face. Although the injuries themselves are rarely directly fatal, serious fights can be very costly, with wounds (mostly on the back or rump) potentially becoming infected, and some defeated rats can even die following a shock or depression-like state. Therefore, serious fighting between familiar individuals is evaded when possible through appeasing body postures, such as lying supine, possibly through audible squeaks and short, variable vocalizations in the 50 kHz range and probably through scent signals, although little is known about these. Rats emit an alarm pheromone in response to aversive events (Burn, 2008), such as electric shocks or predator encounters, and it causes avoidance in conspecifics, but it is not known whether they produce it in defensive contexts. They also produce a porphyrin-rich secretion, chromodacryorrhoea (or 'red tears'), from the Harderian glands behind the eyes in response to diverse stressors, but it is unclear whether chromodacryorrhoea has an appeasing effect in the Norway rat. The vibrissae (whiskers) are very important for preventing bites to the face, and vibrissal contact alone can be sufficient to drive an opponent away (Blanchard *et al.*, 2001).

More is known about aggression than about other social behaviours, but rats actively seek contact with conspecifics. They often rest huddled together in a nest cavity (Calhoun, 1963) and, in captivity at least, are among the few species that occasionally engage in play-like behaviour, even as adults. Play superficially resembles aggression, but the bites cause no injury and are directed at the neck rather than the rump, and the hair does not become pilo-erected. Play sessions often end with one rat being pinned down by the other, who then vigorously nibbles and scratches at the pinned rat's fur: an activity known as allogrooming. The social implications of allogrooming are not well understood.

16.6 Mate Choice, Mating Behaviour and Parental Care

The mouse

Mice are prolific breeders when food and nesting material are abundant. The pest control literature estimates that, in resource-rich commensal environments, females can produce 50 (or more) pups per year (Randall, 1999): indeed, this high reproductive rate is one of the reasons for the success of mice in both natural and laboratory environments. Females have a number of strategies for maximizing their reproductive output. For example, they prefer unfamiliar males that exhibit dominant behaviours, such as strong over-marking of urine marks and high rates of ultrasonic calling (Latham and Mason, 2004). Males similarly also show a preference for unfamiliar females, and for more 'feminine' characteristics, such as a shorter anogenital distance – an indication of reduced prenatal exposure to testosterone. Olfactory cues play a key role in identifying good-quality mates of both sexes, but they also influence the reproductive state in females, for example by suppressing oestrus or by delaying/bringing forward its onset. Females indicate sexual receptivity through hopping and darting movements, ear-wiggling and ultrasonic vocalizations. The male responds with his own ultrasonic vocalizations (sometimes called 'songs') and then mounts and copulates with the female – although subordinate males may not progress with copulation if a more dominant animal appears (Brown, 1953).

Gestation lasts between 19 and 21 days (Berry, 1970) and, during the final few days, females are highly motivated to build complex nests (Brown, 1953; Meehan, 1984). The pups are often born at night, and the female spends time between each birth cleaning the pups and moving around the nest. Litters consist on average of five to nine individuals, with equal numbers of males and females. The newborn mice are blind, hairless and dependent upon their mothers for food and thermoregulation. They elicit parental behaviours such as nursing, grooming and retrieval through specific vocalizations, including ultrasonic calls. In some populations, reproductive females may nest and even nurse their young communally – particularly if the females cannot distinguish their pups from those of other mothers, as may occur if the pups are born within a few days of each other. It is unknown whether communal nursing is an adaptive strategy or a response to suboptimal conditions, but studies have shown it to be beneficial, in terms of both pup survival and pup growth prior to weaning.

The nature and quality of parental care that pups receive, and other aspects of their early (even prenatal) environment, have a strong influence on the pups' development. Thus, for example, maternal stress can slow postnatal growth in the pups, increase their stress responses later in life and reduce territorial and reproductive success. The amount of milk that pups receive can influence when they reach sexual maturity and their adult weight. The amount of licking and grooming that pups receive influences their later stress responses, and this also has a knock-on effect on subsequent generations because it influences how much females lick their own pups once they themselves become mothers. This remarkable developmental plasticity during these early weeks moulds pups so that they are best suited to the environments in which they will live as adults (see Fig. 16.2).

Before weaning (and particularly during the earliest days) the pups are at risk of infanticide, both by the mother (perhaps to optimize litter size) and by males – although the risk of males destroying litters is reduced if the male has mated and subsequently cohabited with a pregnant female. However, those pups that survive grow and develop rapidly, and within 3–4 weeks they have fur and can thermoregulate, their eyes have opened and their vision is developed, they have started eating solid food and are regularly venturing out of, and increasingly away from, the nest (Latham and Mason, 2004). By 5–6 weeks the young mice begin to attain sexual maturity – although this may be delayed to 12 weeks of age if breeding conditions are unfavourable (Van Zeegeren, 1980). The onset of sexual maturity signals a change in family relationships, particularly among the males, where aggression escalates between the breeding male and his sons (Berry, 1970; Bronson, 1979). Most juvenile offspring leave their natal territory at this point and disperse to find their own. However, some offspring remain with their parents where they may 'tough it out' as subordinates, with the possibility that one day they may replace their parent as the breeding male/female (Berry, 1970; Van Zeegeren, 1980).

The rat

Puberty occurs at around 50–60 days in laboratory rats, but in wild rats is probably delayed if resources are scarce. Matings usually occur around dusk. Dominant males

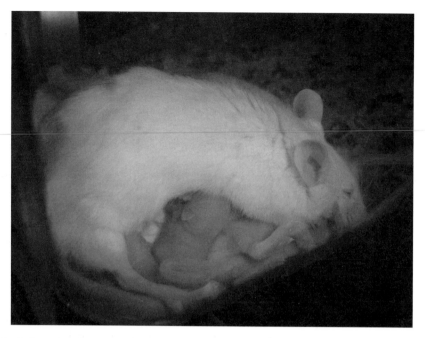

Fig. 16.2. In both mice and rats, higher levels of licking/grooming and active nursing result in lower stress reactivity and fearfulness in the offspring. This also has a knock-on effect on subsequent generations, because it influences the degree to which females lick their own pups once they become mothers. This is a remarkable example of developmental plasticity by which the pups are thought to adjust their defensive neural systems to the demands of the environments in which they will live as adults (image courtesy of Ivana D'Andrea).

mate more often than subordinates, especially in small rat colonies where it is easier for them to monopolize females (Hanson, 2003). Females are fertile only when in oestrus, which occurs every 4–5 days (if they do not conceive) for a period of about 6 h, during which time they may mate repeatedly. Female fecundity is greatest in small, single-male colonies. However, in large colonies many males mate repeatedly with a female, showing little active competition with each other, which causes her stress and reduces her individual fecundity. Given the opportunity, the female will hide in her burrow for a portion of oestrus, and she may turn and box with males to resist the matings.

Nevertheless, oestrous females are proactive participants in sexual encounters, and are attracted to males at least partly through a sex pheromone from the male preputial gland, and volatile male odours advance the onset of puberty and stimulate oestrus in females (Burn, 2008). Sexual receptivity is also signalled largely through scents. Females signal their receptivity by presenting to the chosen male(s), hopping, darting and vibrating their ears, as well as uttering ultrasonic vocalizations. The male also vocalizes and may chase the female. Finally the male mounts the female, who adopts a lordosis posture. After ejaculation, the male stands still and emits a 21–30 kHz vocalization for around 3 min; the pitch and temporal characteristics of this vocalization closely resemble that made in aversive contexts. Gestation lasts for about 21–25 days in the wild. In larger colonies, pups may be sired by several different

H. Würbel *et al.*

fathers, usually the more dominant males (Hanson, 2003). Male rats seem equally fecund all year round, but most pregnancies are observed in spring and autumn when resources are plentiful; only large females can breed year-round (Davis, 1953).

Females sometimes construct new burrows just before giving birth, and will construct more elaborate nests than those in which the adults sleep. In the laboratory at least, parturition can take around 6 h. Several females may nest cooperatively, which usually enhances pup survival (Calhoun, 1963). The normal seven to eight newborn pups in a litter (Davis, 1953) are hairless, blind and ectothermic, relying on their mother's body warmth to maintain their optimal temperature of around 35°C (Alberts, 2005). Olfactory cues on the mother's belly attract pups to suckle, and pheromones deposited into the nesting material reduce their activity levels, keeping them in the nest (Burn, 2008). If they stray from the nest, they therefore become more active; they also emit 40–50 kHz ultrasonic vocalizations, particularly if they become cooled, and these vocalizations cause their mother to retrieve and return them to the nest. Dodecyl propionate from the pup preputial gland stimulates anogenital licking by the mother, which is essential to the pups' survival.

Pregnant and lactating females deposit scents in the nest, preventing cohabiting males from killing the offspring, but intruding males may kill the pups. Males may kill pups that are not their own but, under stable conditions, this is rare because in small colonies almost all pups are sired by the dominant male, and in larger colonies the males cannot usually discriminate which pups are their own because of the polygynandrous mating system (Berdoy, 2002). If resources are scarce or the pups' survival is threatened, the female may kill and/or eat the young, particularly up to about 3 days following birth. This is more likely to occur in subordinate females in poor condition than in healthy, dominant females.

The main developmental stages are as follows if resources are plentiful, but if conditions are suboptimal they may be delayed. The fur appears from about 5 days of age, and pups start walking from about day 10 (Alberts, 2005). The pups suckle for approximately 10 h/day, mostly during the light period, and the dam's milk production reaches a peak around day 15. While some food preferences are established *in utero* (with pups preferring novel foods that the mother consumed during pregnancy), the pups also learn food preferences in the nest from the taste of their mother's milk, olfactory cues on the mother's body and from olfactory and gustatory cues and pheromones in her faeces. From about day 16 onwards the pups may emerge from the nest, and are therefore attracted to these familiar food types if available outside the nest.

Wild female rats have approximately five litters per year, successfully weaning 20–30 pups (Davis, 1953). In the wild, weaning is a gradual process, with suckling ceasing any time between days 16 and 34 or so (Alberts, 2005). As well as being attracted to familiar foods, the weanlings are also attracted to foods that colony members are currently eating, through olfactory cues deposited around the food. Prepubertal rats are not attacked by adults, so they can easily access communal food (Calhoun, 1963). Thus, they avoid consuming potentially toxic substances while sampling a variety of the available foods.

Play in juvenile wild rats is probably important for learning foraging techniques, developing normal social behaviour and establishing dominance hierarchies. Laboratory studies show that social play primarily reflects adult precopulatory behaviours rather than aggressive behaviours. This is indicated by the target of playful attack being the nape of the neck, as in precopulatory behaviour, rather than the rump, which is the usual target for aggressive biting. Mounting and lordosis are rarely seen during play until the rats approach puberty at around 6–8 weeks old.

After puberty males are more likely to disperse than females but, unless the environment is unfavourable, dispersal is rare, with studies showing that fewer than 10% of a colony disperse per year (Davis, 1953). In the wild, most rats that do disperse are unlikely to survive and rarely integrate successfully into existing established colonies. Taken overall, females live about 20% longer than the average male but, even within stable colonies, only about 5% of rats live beyond 1 year.

16.7   Housing and Welfare

Practical and economic considerations mean that housing conditions for laboratory rodents differ dramatically from the natural environments to which they are adapted. It is often argued that hundreds of generations of laboratory breeding must have resulted in animals that are well adapted to these conditions. However, this is supported neither by studies of feral animals or laboratory animals released in wild-type environments (e.g. Berdoy, 2002), nor by studies on the behaviour and welfare of mice or rats under laboratory conditions (Würbel, 2001). Both laboratory mice and rats have retained the behavioural repertoire of their wild ancestors. Although quantitative behavioural changes have occurred (e.g. laboratory breeds are generally more docile), no qualitative differences are detectable. Their behavioural needs seem to have changed little in the laboratory.

This has numerous implications for the welfare of laboratory mice and rats, as well as for the validity of research conducted with them (Würbel, 2001; Latham and Mason, 2004; Sherwin, 2004). However, the relationship between animal welfare and research validity is much more complex than often suggested. On the one hand, research outcomes may be directly compromised by adverse effects of housing on the expression of physiological and behavioural traits (Würbel, 2001), regardless of whether this is associated with impaired well-being. Conversely, poor well-being may or may not impinge on research outcomes, depending on whether it is associated with related behavioural or physiological changes.

In the following sections, we briefly discuss five aspects of laboratory housing which are likely to account for the majority of welfare problems induced by laboratory housing of mice and rats. These five aspects are not independent, however, as they partly overlap in their effects on the animals and may interact with each other to generate further welfare problems.

16.8   Sensory and Motor Deprivation

Standard cages for laboratory rodents are severely restricted in space and impoverished in social and environmental complexity as compared with the animals' natural habitat. In addition the conditions are highly stable, providing little novel stimulation. Laboratory housing conditions therefore induce sensory and motor deprivation (Würbel, 2001), and this is further exacerbated by the current trend towards individually ventilated cage (IVC) systems. For example, mice and rats are fed concen-

Fig. 16.3. In the laboratory, rich sensory and motor stimulation and critical resources may be provided by adequate environmental enrichment, even without interfering with the animals' visibility. (Reprinted from Marashi *et al.*, 2003, with kind permission from Elsevier.)

trated food pellets in hoppers, eliminating the need for extensive, time-consuming foraging and feeding behaviour. Sensory and motor deprivation during development can result in impaired brain and behavioural development (van Praag *et al.*, 2000). Barren environments mainly affect the hippocampus and neocortex, resulting in poor learning and memory (van Praag *et al.*, 2000). In contrast, social deprivation, especially during adolescence, mainly affects the frontal–cortico–striatal pathways, resulting in impaired inhibitory control of behaviour as indicated, e.g. by hyperactivity, stereotypy, compulsive behaviour and inflexibility in higher-order cognitive functions (Würbel, 2001). Taken together, sensory and motor deprivation during development may adversely affect the animals' ability to adjust behaviour rapidly and flexibly to changing conditions. This may seriously compromise their ability to cope with the demands of life as laboratory animals (Würbel, 2001; Fig. 16.3).

16.9 Lack of Critical Resources

Standard cages for laboratory rodents also lack some essential resources and prevent the animals from performing some natural behavioural responses to external stimuli. For example, both mice and rats are highly thigmotactic and, when faced with predator cues, rapidly seek shelter. Although there may be no real predators in a laboratory setting, some cues may still be perceived as predator cues by the animals (Burn, 2008), but often no shelter is provided for them in which to hide. They also perceive visual, auditory and olfactory cues of conspecifics in neighbouring cages, but are chronically prevented from exploring them properly. In response to the chronic thwarting of highly motivated behaviours, the animals may get stuck in appetitive loops of behaviour, since the consummatory phase is never reached. Over time, the repeated attempts to perform particular behaviours or to reach particular goals may

Fig. 16.4. Mice and other rodents readily develop abnormal stereotypic behaviours when housed in barren laboratory cages. Stereotypic jumping (white mouse to the left and nude mouse to the right) and bar mouthing (white mouse in the right-hand corner of the cage to the left) originate from two different appetitive behaviours to escape from the home cage (image courtesy of Hanno Würbel).

develop into abnormal stereotypic behaviour (Würbel, 2006). Especially mice develop diverse abnormal stereotypic behaviours (e.g. bar mouthing, jumping, somersaulting, pacing, circling, etc.; see Fig. 16.4), most of which seem to originate from attempts to escape from the cage (Würbel, 2006). Such escape attempts may not simply reflect aversion towards the cage generally, but may be more specifically elicited by the lack of some resource (e.g. shelter, nesting material) or attraction towards social cues from neighbouring cages.

However, not all species are similarly prone to stereotypy development. Indeed, while most laboratory mice develop stereotypies at high intensities, rats generally show little stereotypy despite being housed under seemingly similar conditions (Würbel, 2006). The reasons for this species difference have remained elusive. One possibility is that rat cages are smaller relative to body size and therefore may provide insufficient space for the animals to develop elaborate stereotypies. Alternatively, rats may employ more passive coping styles, as indicated by their high levels of inactivity under standard laboratory housing conditions.

16.10 Phenotypic Mismatch and Developmental Disruption

Artificial social and/or environmental conditions may also alter or disrupt adaptive mechanisms of developmental plasticity. Developmental plasticity provides for the fine-tuning of genetically determined response rules. The mechanisms underlying developmental plasticity are generally triggered by early environmental or social cues (e.g. predator cues, maternal care) that are somehow predictive of the animals' future environment, allowing them to adjust their phenotype to the specific demands of that environment. Under laboratory conditions, however, the outcome may be maladaptive if these early cues do not properly predict the animals' future environment, resulting in phenotypic mismatch or developmental disruptions (Macri and Würbel, 2006). For example, in rats (and possibly mice), environmentally dependent variations in maternal care play a significant role in 'programming' the pups' later responses to stressors (Meaney, 2001). Under laboratory conditions, mice and rats are normally bred under conditions of minimal disturbance and interference in view of maximizing breeding success. However, these conditions are associated with low levels of active maternal care, resulting in offspring that are highly vulnerable to stressors later in life (Macri and Würbel, 2006). Thus, standard laboratory breeding conditions may produce a phenotypic mismatch, resulting in animals that are poorly equipped to cope with the challenges of life as a laboratory animal.

16.11 Social Conflict

Unstable social groups may induce social conflict, leading to aggression, social stress and injuries that may even cause death, especially in mice. This is further worsened by the fact that some naturally variable aspects of the social environment, e.g. individual odour cues, are unnaturally homogeneous due to inbreeding, while other naturally stable aspects of the environment, e.g. the build-up of odour cues, are regularly disrupted (e.g. during cage cleaning). Furthermore, in male mice social conflict and overt aggression may be increased by enrichments, especially when they can be monopolized. Thus, although the addition of critical resources such as shelters would seem to benefit the animals as discussed above, they may induce competition leading to aggression. This should not be taken as an argument against providing these resources. However, it demonstrates how the different aspects listed here may interact, and highlights the difficulties in designing housing conditions in a way that integrates these different aspects into a functioning system. Although overt aggression appears to be less of a problem in laboratory rats, social stress as a result of non-harmonious groups can occur. This suggests that aggression should be monitored and individuals separated from non-harmonious groups.

16.12 Environmental Stressors

Finally, laboratory animals are generally exposed to a multitude of environmental stressors, many of which are unpredictable. The sensory environment itself may cause stress, with the animals being inescapably exposed to high artificial light intensities, audible and ultrasonic sounds that can be very loud, vibration (especially in IVCs) and salient odours that can potentially include other laboratory animals (even non-rodent species), cleaning products and myriad other chemicals (Latham and Mason, 2004; Burn, 2008). Normal husbandry events, such as handling and cage cleaning, and procedures specific to individual experimental purposes can cause further stress. In addition, the confines of barren cages provide the animals with little control over these stressors. Predictability and controllability have long been shown to be the major determinants of the stress response. Moreover, because many stressful events occur in the rodents' resting phase (when it is light), the animals are less able to cope with them than if they occur during their active phase. Consequently, it is not surprising to find evidence for stress- and anxiety-related disorders (e.g. compulsive behaviours such as barbering in mice) and depressive-like states (e.g. anhedonia – the inability to feel pleasure – in rats).

16.13 Conclusions

Despite hundreds of generations of laboratory breeding, mice and rats have retained their behavioural repertoire. Although quantitative behavioural changes have occurred, the animals' basic behavioural needs and their natural response rules seem to have remained largely unchanged by domestication in the laboratory. Laboratory conditions that ignore the animals' ethology are therefore likely to induce poor welfare and compromise the validity of animal experiments.

By understanding the ethology of laboratory animals we can gain insight into how the captive environments may impact upon animal welfare, and formulate testable hypotheses about potential refinements to laboratory housing and husbandry practices (Olsson and Dahlborn, 2002; Olsson et al., 2003). After all, a laboratory mouse is still a house mouse, and a laboratory rat is still a Norway rat.

References

Alberts, J.A. (2005) Infancy. In: Whishaw, I.Q. and Kolb, B. (eds) *The Behavior of the Laboratory Rat: a Handbook with Tests*. Oxford University Press, Oxford, UK, pp. 266–277.

Barnett, S.A. (2005) Ecology. In: Whishaw, I.Q. and Kolb, B. (eds) *The Behavior of the Laboratory Rat: a Handbook with Tests*. Oxford University Press, Oxford, UK, pp. 15–24.

Baumgardner, D., Ward, S. and Dewsbury, D. (1980) Diurnal patterning of eight activities in 14 species of muroid rodent. *Animal Learning and Behaviour* 8, 322–330.

Berdoy, M. (2002) The laboratory rat: a natural history (accessed 24 Feb. 2006 from http://www.ratlife.org).

Berry, R. (1970) The natural history of the house mouse. *Field Studies* 3, 219–262.

Blanchard, R.J., Dulloog, L., Markham, C., Nishimura, O., Nikulina Compton, J., Jun, A., Han, C. and Blanchard, D.C. (2001) Sexual and aggressive interactions in a visible burrow system with provisioned burrows. *Physiology and Behavior* 72, 245–254.

Bronson, F. (1979) The reproductive ecology of the house mouse. *The Quarterly Review of Biology* 54, 265–299.

Brown, R. (1953) Social behavior, reproduction and population changes in the house mouse (*Mus musculus* L.). *Ecological Monographs* 23, 217–240.

Burn, C.C. (2008) What is it like to be a rat? Rat sensory perception and its implications for experimental design and rat welfare. *Applied Animal Behaviour Science* 112, 1–32.

Calhoun, J.B. (1963) *The Ecology and Sociology of the Norway Rat*. Vol. 1008, US Department of Health Education and Welfare Public Health Service, Bethesda, Maryland.

Crawley, J. (2000) *What's Wrong with my Mouse? Behavioural Phenotyping of Transgenic and Knockout Mice*. Wiley-Liss and Sons, Inc., Chichester, UK.

Crowcroft, P. (1966) *Mice All Over*. Foulis, London.

Davis, D.E. (1953) The characteristics of rat populations. *The Quarterly Review of Biology* 28, 373.

Galef, B.G., Jr., Mason, J.R., Preti, G. and Bean, N.J. (1988) Carbon disulfide: a semiochemical mediating socially induced diet choice in rats. *Physiology and Behavior* 42, 119–124.

Hanson, A. (2003) *Rat Behaviour and Biology* (accessed 22 March 2007 from http://www.ratbehavior.org).

Hurst, J., Payne, C., Nevison, C., Marie, A., Humphries, R., Robertson, D., Cavaggioni, A. and Beynon, R. (2001) Individual recognition in mice mediated by major urinary proteins. *Nature* 414, 631–634.

Latham, N. and Mason, G. (2004) From house mouse to mouse house: the behavioural biology of free-living *Mus musculus* and its implications in the laboratory. *Applied Animal Behaviour Science* 86, 261–289.

Macrì, S. and Würbel, H. (2006) Developmental plasticity of HPA and fear responses in rats: a critical review of the maternal mediation hypothesis. *Hormones and Behavior* 50, 667–680.

Meaney, M.J. (2001) Maternal care, gene expression, and the transmission of individual differences in stress reactivity across generations. *Annual Review of Neuroscience* 24, 1161–1192.

Meehan, A. (1984) *Rats and Mice: Their Biology and Control.* Brown, Knight and Truscott Ltd., Tonbridge, UK.

Olsson, A. and Dahlborn, K. (2002) Improving housing conditions for laboratory mice: a review of 'environmental enrichment'. *Laboratory Animals* 36, 243–270.

Olsson, I.A.S., Nevison, C.M., Patterson-Kane, E.G., Sherwin, C.M., Van de Weerd, H.A. and Würbel, H. (2003) Understanding behaviour: the relevance of ethological approaches in laboratory animal science. *Applied Animal Behaviour Science* 81, 243–264.

Randall, C. (1999) *Vertebrate Pest Management – a Guide for Commercial Applicators* (Extension Bulletin E-2050). Michigan State University, East Lansing, Michigan.

Sherwin, C.M. (2004) The influences of standard laboratory cages on rodents and the scientific validity of research data. *Animal Welfare* 13, S9–S15.

Traweger, D., Travnitzky, R., Moser, C., Walzer, C. and Bernatzky, G. (2006) Habitat preferences and distribution of the brown rat (*Rattus norvegicus* Berk.) in the city of Salzburg (Austria): implications for urban rat management. *Journal of Pest Science* 79, 113–125.

van Praag, H., Kempermann, G. and Gage, F.H. (2000) Neural consequences of environmental enrichment. *Nature Reviews. Neuroscience* 1, 191–198.

Van Zeegeren, K. (1980) Variation in aggressiveness and the regulation of numbers in house mouse populations. *The Netherlands Journal of Zoology* 30, 635–770.

Würbel, H. (2001) Ideal homes? Housing effects on rodent brain and behaviour. *Trends in Neuroscience* 24, 207–211.

Würbel, H. (2006) The motivational basis of caged rodents' stereotypies. In: Mason, G.J. and Rushen, J. (eds) *Stereotypic Animal Behaviour – Fundamentals and Applications to Welfare.* CAB International, Wallingford, UK, pp. 86–120.

Index

Page numbers in **bold** refer to illustrations and tables

Breeding (*continued*)
    purposes, fowl 121
    seasonality 80
    see also Reproduction
British Farm Animal Welfare
        Council 94
Browsing 140, 166

Cannibalism 88–89, 125–126, 227
    see also Infanticide
Cannon, Walter B. 89
Categorization 65–66, 69
Cats
    behaviour development 211–212
    breeds 204–205
    communication 117, 206–208
    domestication 204–205
    dominance 210–211
    human interactions 111, 115, 214
    hunting 205–206
    origins 204–205
    paganism association 204
    play 212, 213
    reproduction 209–**210**
    signals 207–209, 211
    social behaviour 206–209
    socialization 111, 112, 114, 118,
        212–214
    vision 211–212
Cattle
    breeds 151–152
    diurnal rhythms 156
    domestication 151
    dominance hierarchies 154
    feeding **153**, 154–**155**, 156, 159
    foraging 154–155
    housing systems **152**, **153**, 156
    libido assessment 157
    lying patterns **155**
    mating 156–157
    nursing 158–159
    offspring care 158–160
    offspring care and development
        158–160
    origins 151–152
    parturition 157–158
    reproduction 156
    sexual behaviour 156–157
    shade access 156
    social behaviour 152–154
    weather effect 156

Causation
    aggression 8, 154
    behaviour 6, 9, 18
    mechanism 55
    motivational 38, **39**–40
    see also Motivation
Chickens *see* Poultry
Climate 156, 163, 165, 168, 175
Cognition
    ability 180, 181–182
    associative learning 60–61
    Cognitive Dysfunction Syndrome 199
    defined 6
    emotional state indicator 68–70
    feeding pleasure 33
    instincts 4, 10
    judgement bias task **69**
    processes 57
    social 66–68
Communication
    before hatching 134
    benefits and costs 75
    human–animal 111, 116–117
    methods 74–**75**, 196–199, 206–209
    olfactory cues 180, 181, 198, 220, 223
    sexual 142–143
    signals 124
    signals comprehension ability 117
    social cognition 66–67
    see also Cues; Signals
Competition 77–79, 163–164
    see also Dominance; Peck order
Conditioning 58–59
    see also Programming; Socialization;
        Training
Conflict, social 231
    see also Dominance; Fighting
Consciousness 27
Consumer demand theory 99–**100**
Consummation 46–47
Control 8, 91
Cooperation 206–207, 213–214, 227
    see also Synchrony
Coping 92, 183–184
Copulation 143, 186
    see also Reproduction
Copying 124–125
    see also Imprinting; Learning
Costs *see* Benefits and costs
Courtship
    cues 42
    displays 18–**19**, **42**, 209

Olfaction (*continued*)
  odour perception 207–208
  protein synthesis role 15
  recognition cues 172, 180, 223
  sexual behaviour role 143
  suckling cues 227
  urinary odour 198, 207, 220–221,
    222–223
  see also Cues; Scent
Ontogeny 6, 9, 55, 211–212
Opioids, endogenous 28–29
Organization 41–47, 222–224
Ostrich, *Struthio camelus* 122, 123
Oxytocin 35, 75, 187

Parent–offspring interaction 80–82,
  134, 171
Parturition 144–145, 157–158, 170,
  186, 187, 200
Peacock 18–**19**
Peck order 78, **79**, **124**, 194, 210
Pecking 88, 125
Perception 110, 180–182, 207–208
Periparturation 186–187
Personalities 114, 182, 214
Phenotypes 15–16, 230–231
Pheromones 224, 226
Phylogeny 6, 9, 18, 137
Physiology 25–36, **90**, 92–93, 140,
  165–166
  see also Brain; Glands
Pigs
  ancestral behaviour patterns 23
  back to nature 85, **86**
  behavioural development 188–190
  breeds 177
  cleanliness 183, 184
  cognition 180, 181–182
  communication 181, 182
  domestication 21, **23**, 177
  dominance hierarchies 178–179
  environment coping 183–184
  feeding 182–183
  foraging 182–183
  lactation behaviour 187–188
  learning and cognition 57
  mating 184–186
  origins 177–178
  parental–offspring interactions 81, 187
  perception 180–181
  reproduction 34–**35**, 186–187

  selection 177, 187
  social behaviour 178–180
  thermoneutrality 183-**184**
Placentophagia 157–158
Plasticity development 230
Play
  aggression resemblance 224
  barren environment effect 189–190
  behaviour development 135, 173–174,
    212
  function 82–83, 227–228
  geldings 142
  human–animal relationship
    importance **113**
  imprint training 146
  object 205–206, 212
  social 213
  stimulation 189
  welfare indicator 189
Poultry
  ancestral behaviour patterns 23
  communication signals 67, 124
  domesticated species 121–122, 130
  dominance hierarchies 124
  egg laying:incubation:hatching
    behaviour 129, 130, 133–134
  energetically costly behaviour
    strategies reduction 23
  feeding 122, 126–127, 130
  foraging 126
  frustration related behaviour 98
  group-living 125
  housing systems 7, 122–**123**
  information integration 60–61
  learning 66, 135
  mating systems 130
  offspring development 134–135
  origins 121
  peck order **79**
  reproduction seasonality 33
  rhythms, diurnal **131**
  roosting behaviour 97, 98
  selection, group-level 77
  sexual behaviour 130–133
  sleep 129
  social behaviour 123–126
  species-typical behaviour patterns 6
  strategy selection self-control 70
  welfare 6–**7**, 95
  see also Birds; Fowl
Predation fear 97, 163, 175, 229
Preening **128**, 129, 130